I0006169

Machine Learning Automation with TPOT

Build, validate, and deploy fully automated machine
learning models with Python

Dario Radečić

BIRMINGHAM—MUMBAI

Machine Learning Automation with TPOT

Group Product Manager: Kunal Parikh
Publishing Product Manager: Ali Abidi
Senior Editor: David Sugarman
Content Development Editor: Joseph Sunil
Technical Editor: Sonam Pandey
Copy Editor: Safis Editing
Project Coordinator: Aparna Nair
Proofreader: Safis Editing
Indexer: Manju Arasan
Production Designer: Shankar Kalbhor

First published: May 2021

Production reference: 1060421

Published by Packt Publishing Ltd.
Livery Place
35 Livery Street
Birmingham
B3 2PB, UK.

ISBN 978-1-80056-788-7

www.packt.com

Contributors

About the author

Dario Radečić is a full-time data scientist at Neos, in Croatia, a part-time data storyteller at Appsilon, in Poland, and a business owner. Dario has a master's degree in data science and years of experience in data science and machine learning, with an emphasis on automated machine learning. He is also a top writer in artificial intelligence on Medium and the owner of a data science blog called Better Data Science.

About the reviewer

Prajjwal is an Electrical Engineering student at Aligarh Muslim University and a member of the AUV-ZHCET club where he works on computer vision. He is the coordinator of the web-dev team at the IEEE student chapter & a mentor at AMU-OSS. Previously, he worked with DeepSource as the developer relation intern for 6 months.

A book is always a collective effort, I owe thanks to Aligarh Muslim University, AUV-ZHCET, DeepSource, my seniors to help me throughout my programming journey. I would also like to show my gratitude to my brother & mentor Roopak & my parents who on every step encouraged me

Table of Contents

Section 2: TPOT – Practical Classification and Regression

2
Deep Dive into TPOT

3
Exploring Regression with TPOT

4
Exploring Classification with TPOT

5
Parallel Training with TPOT and Dask

Section 3: Advanced Examples and Neural Networks in TPOT

6
Getting Started with Deep Learning: Crash Course in Neural Networks

7
Neural Network Classifier with TPOT

8
TPOT Model Deployment

9
Using the Deployed TPOT Model in Production

Other Books You May Enjoy

Index

Preface

The automation of machine learning tasks allows developers more time to focus on the usability and reactivity of the software powered by machine learning models. TPOT is a Python automated machine learning tool used for optimizing machine learning pipelines using genetic programming. Automating machine learning with TPOT enables individuals and companies to develop production-ready machine learning models more cheaply and quickly than with traditional methods.

With this practical guide to AutoML, developers working with Python on machine learning tasks will be able to put their knowledge to work and become productive quickly. You'll adopt a hands-on approach to learning the implementation of AutoML and its associated methodologies. Complete with step-by-step explanations of essential concepts, practical examples, and self-assessment questions, this book will show you how to build automated classification and regression models and compare their performance to custom-built models. As you advance, you'll also develop state-of-the-art models using only a couple of lines of code and see how those models outperform all of your previous models on the same datasets.

By the end of this book, you'll have gained the confidence to implement AutoML techniques in your organization on a production level.

Who this book is for

Data scientists, data analysts, and software developers who are new to machine learning and want to use it in their applications will find this book useful. This book is also for business users looking to automate business tasks with machine learning. Working knowledge of the Python programming language and a beginner-level understanding of machine learning are necessary to get started.

What this book covers

Chapter 1, Machine Learning and the Idea of Automation, covers a brief introduction to machine learning, the difference between classification and regression tasks, an overview of automation and why it is needed, and the machine learning options in the Python ecosystem.

Chapter 2, Deep Dive into TPOT, provides an in-depth overview of what TPOT is and isn't, how it is used to handle automation in machine learning, and what types of tasks it can automate. This chapter also sees you set up the programming environment.

Chapter 3, Exploring Regression with TPOT, covers the use of TPOT for regression tasks. You'll learn how to apply automated algorithms to data and how to explore your datasets.

Chapter 4, Exploring Classification with TPOT, covers the use of TPOT for classification tasks. You'll learn how to perform basic exploratory data analysis, preparation, train automated models, and compare these automated models with default models from scikit-learn.

Chapter 5, Parallel Training with TPOT and Dask, covers the basics of parallel programming with Python and the Dask library. You'll learn how to use Dask to train automated models in a parallel fashion.

Chapter 6, Getting Started with Deep Learning: A Crash Course in Neural Networks, covers the fundamental ideas behind deep learning, such as neurons, layers, activation functions, and artificial neural networks.

Chapter 7, Neural Network Classifier with TPOT, provides a step-by-step guide to implementing a fully automated neural network classifier, dataset exploration, model training, and evaluation.

Chapter 8, TPOT Model Deployment, takes you through a step-by-step guide to model deployment. You'll learn how to use Flask and Flask-RESTful to build a REST API that is then deployed both locally and to AWS.

Chapter 9, Using the Deployed TPOT Model in Production, covers the usage of the deployed model in a notebook environment and in a simple web application.

To get the most out of this book

You will need Python 3.6 or newer installed on your computer. The code for the book was tested on Python 3.8.x, but any version above 3.6 should work fine. The book isn't OS-specific, as all of the code will work independently of the OS. Still, keep in mind that most of the code was initially run on macOS.

Software/hardware covered in the book	OS requirements
Python 3.6+	Windows, macOS, or Linux (Any)

You don't need any premium or licensed software to follow along with this book. Every library is entirely open source.

If you are using the digital version of this book, we advise you to type the code yourself or access the code via the GitHub repository (link available in the next section). Doing so will help you avoid any potential errors related to the copying and pasting of code.

Download the example code files

You can download the example code files for this book from GitHub at `https://github.com/PacktPublishing/Machine-Learning-Automation-with-TPOT`. In case there's an update to the code, it will be updated on the existing GitHub repository.

We also have other code bundles from our rich catalog of books and videos available at `https://github.com/PacktPublishing/`. Check them out!

Download the color images

We also provide a PDF file that has color images of the screenshots/diagrams used in this book. You can download it here: `https://static.packt-cdn.com/downloads/9781800567887_ColorImages.pdf`

Conventions used

There are a number of text conventions used throughout this book.

`Code in text`: Indicates code words in text, database table names, folder names, filenames, file extensions, pathnames, dummy URLs, user input, and Twitter handles. Here is an example: "Mount the downloaded `WebStorm-10*.dmg` disk image file as another disk in your system."

A block of code is set as follows:

```
output = (inputs[0] * weights[0] +
          inputs[1] * weights[1] +
          inputs[2] * weights[2] +
          inputs[3] * weights[3] +
          inputs[4] * weights[4] +
          bias)
output
```

When we wish to draw your attention to a particular part of a code block, the relevant lines or items are set in bold:

```
CPU times: user 26.5 s, sys: 9.7 s, total: 36.2 s
Wall time: 42 s
```

Any command-line input or output is written as follows:

```
pipenv install "dask[complete]"
```

> **Tips or important notes**
> Appear like this.

Get in touch

Feedback from our readers is always welcome.

General feedback: If you have questions about any aspect of this book, mention the book title in the subject of your message and email us at customercare@packtpub.com.

Errata: Although we have taken every care to ensure the accuracy of our content, mistakes do happen. If you have found a mistake in this book, we would be grateful if you would report this to us. Please visit www.packtpub.com/support/errata, selecting your book, clicking on the Errata Submission Form link, and entering the details.

Piracy: If you come across any illegal copies of our works in any form on the Internet, we would be grateful if you would provide us with the location address or website name. Please contact us at copyright@packt.com with a link to the material.

If you are interested in becoming an author: If there is a topic that you have expertise in and you are interested in either writing or contributing to a book, please visit authors.packtpub.com.

Reviews

Please leave a review. Once you have read and used this book, why not leave a review on the site that you purchased it from? Potential readers can then see and use your unbiased opinion to make purchase decisions, we at Packt can understand what you think about our products, and our authors can see your feedback on their book. Thank you!

For more information about Packt, please visit packt.com.

Section 1: Introducing Machine Learning and the Idea of Automation

This section provides a quick revision of machine learning – classification and regression tasks, an overview of automation and why it is needed, and what options are available in the Python ecosystem.

This section contains the following chapter:

- *Chapter 1, Machine Learning and the Idea of Automation*

1
Machine Learning and the Idea of Automation

In this chapter, we'll make a quick revision of the essential machine learning topics. Topics such as supervised machine learning are covered, alongside the basic concepts of regression and classification.

We will understand why machine learning is essential for success in the 21st century from various perspectives: those of students, professionals, and business users, and we will discuss the different types of problems machine learning can solve.

Further, we will introduce the concept of automation and understand how it applies to machine learning tasks. We will go over automation options in the Python ecosystem and compare their pros and cons. We will briefly introduce the **TPOT** library, and discuss its role in the modern-day automation of machine learning.

This chapter will cover the following topics:

- Reviewing the history of machine learning

- Reviewing automation

- Applying automation to machine learning

- Automation options for Python

Technical requirements

To complete this chapter, you only need Python installed, alongside the basic data processing and machine learning libraries, such as `numpy`, `pandas`, `matplotlib`, and `scikit-learn`. You'll learn how to install and configure these in a virtual environment in *Chapter 2, Deep Dive into TPOT*, but let's keep this one easy. These libraries come preinstalled with any Anaconda distribution, so you shouldn't have to worry about it. If you are using raw Python instead of Anaconda, executing this line from the Terminal will install everything needed:

```
> pip install numpy pandas matplotlib scikit-learn
```

Keep in mind it's always a good practice to install libraries in a virtual environment, and you'll learn how to do that shortly.

The code for this chapter can be downloaded here:

```
https://github.com/PacktPublishing/Machine-Learning-
Automation-with-TPOT/tree/main/Chapter01
```

Reviewing the history of machine learning

Just over 25 years ago (1994), a question was asked in an episode of *The Today Show* – *"What is the internet, anyway?"* It's hard to imagine that a couple of decades ago, the general population had difficulty defining what the internet is and how it works. Little did they know that we would have intelligent systems managing themselves only a quarter of a century later, available to the masses.

The concept of machine learning was introduced much earlier in 1949 by Donald Hebb. He presented theories on neuron excitement and communication between neurons (*A Brief History of Machine Learning – DATAVERSITY*, Foote, K., March 26, 2019). He was the first to introduce the concept of artificial neurons, their activation, and their relationships through weights.

In the 1950s, Arthur Samuel developed a computer program for playing checkers. The memory was quite limited at that time, so he designed a scoring function that attempted to measure every player's probability of winning based on the positions on the board. The program chose its next move using a MinMax strategy, which eventually evolved into the MinMax algorithm (*A Brief History of Machine Learning – DATAVERSITY*, Foote, K., March 26; 2019). Samuel was also the first one to come up with the term **machine learning**.

Frank Rosenblatt decided to combine Hebb's artificial brain cell model with the work of Arthur Samuel to create a **perceptron**. In 1957, a perceptron was planned as a machine, which led to building a **Mark 1 perceptron** machine, designed for image classification.

The idea seemed promising, to say at least, but the machine couldn't recognize useful visual patterns, which caused a stall in further research – this period is known as the first AI winter. There wasn't much going on with the perceptron and neural network models until the 1990s.

The preceding couple of paragraphs tell us more than enough about the state of machine learning and deep learning at the end of the 20th century. Groups of individuals were making tremendous progress with neural networks, while the general population had difficulty understanding even what the internet is.

To make machine learning useful in the real world, scientists and researchers required two things:

- **Data**
- **Computing power**

The first was rapidly becoming more available due to the rise of the internet. The second was slowly moving into a phase of exponential growth – both in CPU performance and storage capacity.

Still, the state of machine learning in the late 1990s and early 2000s was nowhere near where it is today. Today's hardware has led to a significant increase in the use of machine-learning-powered systems in production applications. It is difficult to imagine a world where Netflix doesn't recommend movies, or Google doesn't automatically filter spam from regular email.

But, what is machine learning, anyway?

What is machine learning?

There are a lot of definitions of machine learning out there, some more and some less formal. Here are a couple worth mentioning:

- Machine learning is an application of **artificial intelligence** (**AI**) that provides systems the ability to automatically learn and improve from experience without being explicitly programmed (*What is Machine Learning? A Definition – Expert System,* Expert System Team; May 6, 2020).

- Machine learning is the concept that a computer program can learn and adapt to new data without human intervention (*Machine Learning – Investopedia,* Frankenfield, J.; August 31, 2020).

- Machine learning is a field of computer science that aims to teach computers how to learn and act without being explicitly programmed (*Machine Learning – DeepAI,* Deep AI Team; May 17, 2020).

Even though these definitions are expressed differently, they convey the same information. Machine learning aims to develop a system or an algorithm capable of learning from data without human intervention.

The goal of a data scientist isn't to instruct the algorithm on how to learn, but rather to provide an adequately sized and prepared dataset to the algorithm and briefly specify the relationships between the dataset variables. For example, suppose the goal is to produce a model capable of predicting housing prices. In that case, the dataset should provide observations on a lot of historical prices, measured through variables such as location, size, number of rooms, age, whether it has a balcony or a garage, and so on.

It's up to the machine learning algorithm to decide which features are important and which aren't, ergo, which features have significant predictive power. The example in the previous paragraph explained the idea of a **regression** problem solved with **supervised machine learning** methods. We'll soon dive into both concepts, so don't worry if you don't quite understand it.

Further, we might want to build a model that can predict, with a decent amount of confidence, whether a customer is likely to churn (break the contract). Useful features would be the list of services the client is using, how long they have been using the service, whether the previous payments were made on time, and so on. This is another example of a supervised machine learning problem, but the target variable (churn) is categorical (yes or no) and not continuous, as was the case in the previous example. We call these types of problems **classification machine learning problems**.

Machine learning isn't limited to regression and classification. It is applied to many other areas, such as clustering and dimensionality reduction. These fall into the category of **unsupervised machine learning** techniques. These topics won't be discussed in this chapter.

But first, let's answer a question on the usability of machine learning models, and discuss who uses these models and in which circumstances.

In which sectors are the companies using machine learning?

In a single word – everywhere. But you'll have to continue reading to get a complete picture. Machine learning has been adopted in almost every industry in the last decade or two. The main reason is the advancements in hardware. Also, machine learning has become easier for the broader masses to use and understand.

It would be impossible to list every industry in which machine learning is used and to discuss further the specific problems it solves. The easier task would be to list the industries that can't benefit from machine learning, as there are far fewer of those.

We'll focus only on the better-known industries in this section.

Here's a list and explanation of the ten most common use cases of machine learning, both from the industry standpoint and as a general overview:

- **The finance industry**: Machine learning is gaining more and more popularity in the financial sector. Banks and financial institutions can use it to make smarter decisions. With machine learning, banks can detect clients who most likely won't repay their loans. Further, banks can use machine learning methods to track and understand the spending patterns of their customers. This can lead to the creation of more personalized services to the satisfaction of both parties. Machine learning can also be used to detect anomalies and fraud through unexpected behaviors on some client accounts.

- **Medical industry**: The recent advancements in medicine are at least partly due to advancements in machine learning. Various predictive methods can be used to detect diseases in the early stages, based on which medical experts can construct personalized therapy and recovery plans. Computer vision techniques such as image classification and object detection can be used, for example, to perform classification on lung images. These can also be used to detect the presence of a tumor based on a single image or a sequence of images.

- **Image recognition**: This is probably the most widely used application of machine learning because it can be applied in any industry. You can go from a simple cat-versus-dog image classification to classifying the skin conditions of endangered animals in Africa. Image recognition can also be used to detect whether an object of interest is present in the image. For example, the automatic detection of Waldo in the *Where's Waldo?* game has roughly the same logic as an algorithm in autonomous vehicles that detects pedestrians.

- **Speech recognition**: Yet another exciting and promising field. The general idea is that an algorithm can automatically recognize the spoken words in an audio clip and then convert them to a text file. Some of the better-known applications are appliance control (controlling the air conditioner with voice commands), voice dialing (automated recognition of a contact to call just from your voice), and internet search (browsing the web with your voice). These are only a couple of examples that immediately pop into mind. Automatic speech recognition software is challenging to develop. Not all languages are supported, and many non-native speakers have accents when speaking in a foreign language, which the ML algorithm may struggle to recognize.

- **Natural Language Processing** (NLP): Companies in the private sector can benefit tremendously from NLP. For example, a company can use NLP to analyze the sentiments of online reviews left by their customers if there are too many to classify manually. Further, companies can create chatbots on web pages that immediately start conversations with users, which then leads to more potential sales. For a more advanced example, NLP can be used to write summaries of long documents and even segment and analyze protein sequences.

- **Recommender systems**: As of late 2020, it's difficult to imagine a world where Google doesn't tailor the search results based on your past behaviors, Amazon doesn't automatically recommend similar products, Netflix doesn't recommend movies and TV shows based on the past watches, and Spotify doesn't recommend music that somehow flew under your radar. These are only a couple of examples, but it's not difficult to recognize the importance of recommender systems.

- **Spam detection**: Just like it's hard to imagine a world where the search results aren't tailored to your liking, it's also hard to imagine an email service that doesn't automatically filter out messages about that now-or-never discount on a vacuum cleaner. We are bombarded with information every day, and automatic spam detection algorithms can help us focus on what's important.

- **Automated trading**: Even the stock market is moving too fast to fully capture what's happening without automated means. Developing trading bots isn't easy, but machine learning can help you pick the best times to buy or sell, based on tons of historical data. If fully automated, you can watch how your money creates money while sipping margaritas on the beach. It might sound like a stretch to some of you, but with robust models and a ton of domain knowledge, I can't see why not.

- **Anomaly detection**: Let's dial back to our banking industry example. Banks can use anomaly detection algorithms for various use cases, such as flagging suspicious transactions and activities. Lately, I've been using anomaly detection algorithms to detect suspicious behavior in network traffic with the goal of automatic detection of cyberattacks and malware. It is another technique applicable to any industry if the data is formatted in the right way.

- **Social networks**: How many times has Facebook recommended you people you may know? Or YouTube recommended the video on the topic you were just thinking about? No, they are not reading your mind, but they are aware of your past behaviors and decisions and can predict your next move with a decent amount of confidence.

These are just a couple of examples of what machine learning can do – not an exhaustive list by any means. You are now familiar with a brief history of machine learning and know how machine learning can be applied to a wide array of tasks.

The next section will provide a brief refresher on supervised machine learning techniques, such as regression and classification.

Supervised learning

The majority of practical machine learning problems are solved through **supervised learning** algorithms. Supervised learning refers to a situation where you have an input variable (a predictor), typically denoted with X, and an output variable (what you are trying to predict), typically denoted with y.

There's a reason why features (X) are denoted with a capital letter and the target variable (y) isn't. In math terms, X denotes a matrix of features, and matrices are typically denoted with capital letters. On the other hand, y is a vector, and lowercase letters are typically used to denote vectors.

The goal of a supervised machine learning algorithm is to learn the function that can transform any input into the output. The most general math representation of a supervised learning algorithm is represented with the following formula:

$$y = f(X)$$

Figure 1.1 – General supervised learning formula

We must apply one of two corrections to make this formula acceptable. The first one is to replace y with y-hat, as y generally denotes the true value, and y-hat denotes the prediction. The second correction we could make is to add the error term, as only then can we have the correct value of y on the other side. The error term represents the irreducible error – the type of error that can't be reduced by further training.

Here's how the first corrected formula looks:

$$\hat{y} = f(X)$$

Figure 1.2 – Corrected supervised learning formula (v1)

And here's the second one:

$$y = f(X) + \varepsilon$$

Figure 1.3 – Corrected supervised learning formula (v2)

It's more common to see the second one, but don't be confused by any of the formats – these formulas generally represent the same thing.

Supervised machine learning is called "supervised" because we have the labeled data at our disposal. You might have already picked this because of the feature and target discussion. This means that we have the correct answers already, ergo, we know which combinations of X yield the corresponding values of y.

The end goal is to make the best generalization from the data available. We want to produce the most unbiased model capable of generalizing on new, unseen data. The concepts of overfitting, underfitting, and the bias-variance trade-off are important to produce such a model, but they are not in the scope of this book.

As we've already mentioned, supervised learning problems are grouped into two main categories:

- **Regression**: The target variable is continuous in nature, such as the price of a house in USD, the temperature in degrees Fahrenheit, weight in pounds, height in inches, and so on.

- **Classification:** The target variable is a category – either binary (true/false, positive/negative, disease/no disease), or multi-class (no symptoms/mild symptoms/severe symptoms, school grades, and so on).

Both regression and classification are explored in the following sections.

Regression

As briefly discussed in the previous sections, regression refers to a phenomenon where the target variable is continuous. The target variable could represent a price, a weight, or a height, to name a few.

The most common type of regression is **linear regression**, a model where a linear relationship between variables is assumed. Linear regression further divides into a simple linear regression (only one feature), and multiple linear regression (multiple features).

> **Important note**
>
> Keep in mind that linear regression isn't the only type of regression. You can perform regression tasks with algorithms such as decision trees, random forests, support vector machines, gradient boosting, and artificial neural networks, but the same concepts still apply.

To make a quick recap of the regression concept, we'll declare a simple `pandas.DataFrame` object consisting of two columns – `Living area` and `Price`. The goal is to predict the price based only on the living space. We are using a simple linear regression model here just because it makes the data visualization process simpler, which, as the end result, makes the regression concept easy to understand:

1. The following is the dataset – both columns contain arbitrary and made-up values:

```
import pandas as pd

df = pd.DataFrame({
    'LivingArea': [300, 356, 501, 407, 950, 782,
                   664, 456, 673, 821, 1024, 900,
                   512, 551, 510, 625, 718, 850],
    'Price': [100, 120, 180, 152, 320, 260,
              210, 150, 245, 300, 390, 305,
              175, 185, 160, 224, 280, 299]
})
```

2. To visualize these data points, we will use the `matplotlib` library. By default, the library doesn't look very appealing, so a couple of tweaks are made with the `matplotlib.rcParams` package:

```
import matplotlib.pyplot as plt
from matplotlib import rcParams

rcParams['figure.figsize'] = 14, 8
rcParams['axes.spines.top'] = False
rcParams['axes.spines.right'] = False
```

3. The following options make the charts larger by default, and remove the borders (spines) on the top and right. The following code snippet visualizes our dataset as a two-dimensional scatter plot:

```
plt.scatter(df['LivingArea'], df['Price'],
color='#7e7e7e', s=200)
plt.title('Living area vs. Price (000 USD)', size=20)
plt.xlabel('Living area', size=14)
plt.ylabel('Price (000 USD)', size=14)
plt.show()
```

The preceding code produces the following graph:

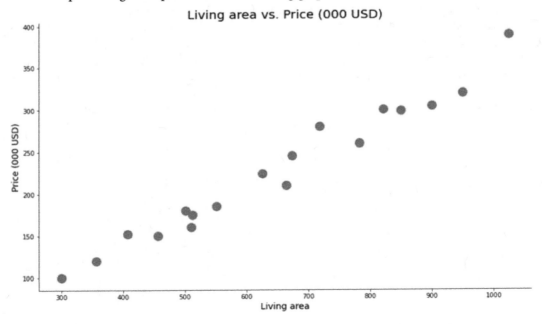

Figure 1.4 – Regression – Scatter plot of living area and price (000 USD)

4. Training a linear regression model is most easily achieved with the `scikit-learn` library. The library contains tons of different algorithms and techniques we can apply on our data. The `sklearn-learn.linear_model` module contains the `LinearRegression` class. We'll use it to train the model on the entire dataset, and then to make predictions on the entire dataset. That's not something you would usually do in production environment, but is essential here to get a further understanding of how the model works:

```
from sklearn.linear_model import LinearRegression
model = LinearRegression()
model.fit(df[['LivingArea']], df[['Price']])

preds = model.predict(df[['LivingArea']])
df['Predicted'] = preds
```

5. We've assigned the prediction as yet another dataset column, just to make data visualization simpler. Once again, we can create a chart containing the entire dataset as a scatter plot. This time, we will add a line that represents the *line of best fit* – the line where the error is smallest:

```
plt.scatter(df['LivingArea'], df['Price'],
color='#7e7e7e', s=200, label='Data points')
plt.plot(df['LivingArea'], df['Predicted'],
color='#040404', label='Best fit line')

plt.title('Living area vs. Price (000 USD)', size=20)
plt.xlabel('Living area', size=14)
plt.ylabel('Price (000 USD)', size=14)
plt.legend()
plt.show()
```

The preceding code produces the following graph:

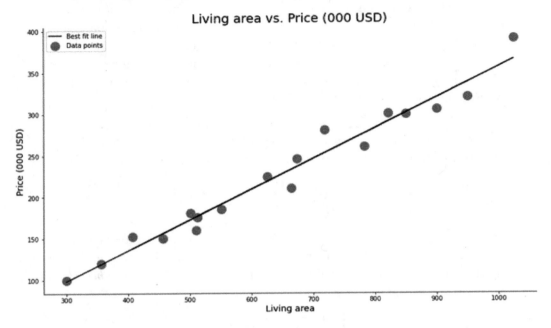

Figure 1.5 – Regression – Scatter plot of living area and price (000 USD) with the line of best fit

6. As we can see, the simple linear regression model almost perfectly captures our dataset. This is not a surprise, as the dataset was created for this purpose. New predictions would be made along the line of best fit. For example, if we were interested in predicting the price of house that has a living space of 1,000 square meters, the model would make a prediction just a bit north of $350K. The implementation of this in the code is simple:

```
model.predict([[1000]])
>>> array([[356.18038708]])
```

7. Further, if you were interested in evaluating this simple linear regression model, metrics like *R2* and *RMSE* are a good choice. R2 measures the goodness of fit, ergo it tells us how much variance our model captures (ranging from 0 to 1). It is more formally referred to as the *coefficient of determination*. RMSE measures how wrong the model is on average, in the unit of interest. For example, an RMSE value of 10 would mean that on average our model is wrong by $10K, in either the positive or negative direction.

Both the R2 score and the RMSE are calculated as follows:

```
import numpy as np
from sklearn.metrics import r2_score, mean_squared_error

rmse = lambda y, ypred: np.sqrt(mean_squared_error(y,
ypred))

model_r2 = r2_score(df['Price'], df['Predicted'])
model_rmse = rmse(df['Price'], df['Predicted'])
print(f'R2 score: {model_r2:.2f}')
print(f'RMSE: {model_rmse:.2f}')

>>> R2 score: 0.97
>>> RMSE: 13.88
```

To conclude, we've built a simple but accurate model. Don't expect data in the real world to behave this nicely, and also don't expect to build such accurate models most of the time. The process of model selection and tuning is tedious and prone to human error, and that's where automation libraries such as **TPOT** come into play.

We'll cover a classification refresher in the next section, again on the fairly simple example.

Classification

Classification in machine learning refers to a type of problem where the target variable is categorical. We could turn the example from the *Regression* section in the classification problem by converting the target variable into categories, such as *Sold/Did not sell*.

In a nutshell, classification algorithms help us in various scenarios, such as predicting customer attrition, whether a tumor is malignant or not, whether someone has a given disease or not, and so on. You get the point.

Classification tasks can be further divided into binary classification tasks and multi-class classification tasks. We'll explore binary classification tasks briefly in this section. The most basic classification algorithm is *logistic regression*, and we'll use it in this section to build a simple classifier.

> **Note**
>
> Keep in mind that you are not limited only to logistic regression for performing classification tasks. On the contrary – it's good practice to use a logistic regression model as a baseline, and to use more sophisticated algorithms in production. More sophisticated algorithms include decision trees, random forests, gradient boosting, and artificial neural networks.

The data is completely made up and arbitrary in this example:

1. We have two columns – the first indicates a measurement of some sort
    (called `Radius`), and the second column denotes the classification (either 0 or 1).
    The dataset is constructed with the following Python code:

    ```python
    import numpy as np
    import pandas as pd


    df = pd.DataFrame({
        'Radius': [0.3, 0.1, 1.7, 0.4, 1.9, 2.1, 0.25,
                   0.4, 2.0, 1.5, 0.6, 0.5, 1.8, 0.25],
        'Class': [0, 0, 1, 0, 1, 1, 0,
                  0, 1, 1, 0, 0, 1, 0]
    })
    ```

2. We'll use the `matplotlib` library once again for visualization purposes. Here's
    how to import it and make it a bit more visually appealing:

    ```python
    import matplotlib.pyplot as plt
    from matplotlib import rcParams

    rcParams['figure.figsize'] = 14, 8
    rcParams['axes.spines.top'] = False
    rcParams['axes.spines.right'] = False
    ```

3. We can reuse the same logic from the previous regression example to make
    a visualization. This time, however, we won't see data that closely resembles a line.
    Instead, we'll see data points separated into two groups. On the lower left are the
    data points where the `Class` attribute is 0, and on the right where it's 1:

    ```python
    plt.scatter(df['Radius'], df['Class'], color='#7e7e7e',
    s=200)

    plt.title('Radius classification', size=20)
    plt.xlabel('Radius (cm)', size=14)
    plt.ylabel('Class', size=14)
    plt.show()
    ```

The following graph is the output of the preceding code:

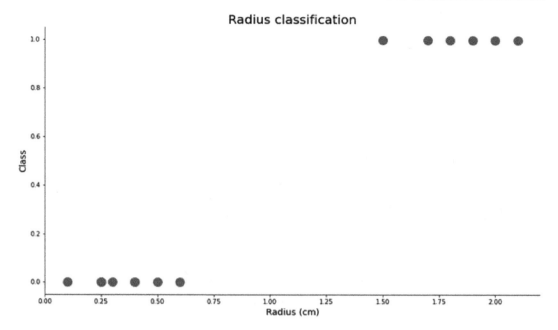

Figure 1.6 – Classification – Scatter plot between measurements and classes

The goal of a classification model isn't to produce a line of best fit, but instead to draw out the best possible separation between the classes.

4. The logistic regression model is available in the sklearn.linear_model package. We'll use it to train the model on the entire dataset, and then to make predictions on the entire dataset. Again, that's not something we will keep doing later on in the book, but is essential to get insights into the inner workings of the model at this point:

```
from sklearn.linear_model import LogisticRegression

model = LogisticRegression()
model.fit(df[['Radius']], df['Class'])

preds = model.predict(df[['Radius']])
df['Predicted'] = preds
```

5. We can now use this model to make predictions on an arbitrary number of X values, ranging from the smallest to the largest in the entire dataset. The range of evenly spaced numbers is obtained through the np.linspace method. It takes three arguments – start, stop, and the number of elements. We'll set the number of elements to 1000.

6. Then, we can make a line that indicates the probabilities for every value of X generated. By doing so, we can visualize the decision boundary of the model:

```
xs = np.linspace(0, df['Radius'].max() + 0.1, 1000)
ys = [model.predict([[x]]) for x in xs]

plt.scatter(df['Radius'], df['Class'], color='#7e7e7e',
s=200, label='Data points')
plt.plot(xs, ys, color='#040404', label='Decision
boundary')

plt.title('Radius classification', size=20)
plt.xlabel('Radius (cm)', size=14)
plt.ylabel('Class', size=14)
plt.legend()
plt.show()
```

The preceding code produces the following visualization:

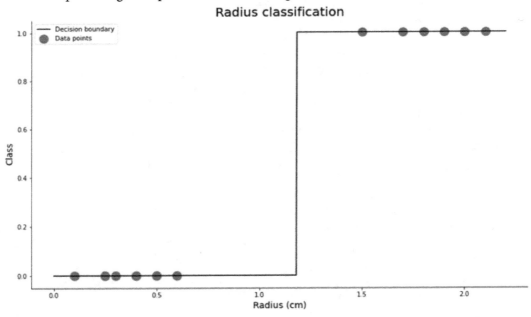

Figure 1.7 – Classification – Scatter plot between measurements and classes and the decision boundary

Our classification model is basically a step function, which is understandable for this simple problem. Nothing more complex is needed to correctly classify every instance in our dataset. This won't always be the case, but more on that later.

7. A confusion matrix is one of the best methods for evaluating classification models. Our *negative* class is 0, and the *positive* class is 1. The confusion matrix is just a square matrix that shows the following:

- *True negatives*: The upper left number. These are instances that had the class of 0 and were predicted as 0 by the model.

- *False negatives*: The bottom left number. These are instances that had the class of 0, but were predicted as 1 by the model.

- *False positives*: The top right number. These are instances that had the class of 1, but were predicted as 0 by the model.

- *True positives*: The bottom right number. These are instances that had the class of 1 and were predicted as 1 by the model.

   Read the previous list as many times as necessary to completely understand the idea. The confusion matrix is an essential concept in classifier evaluation, and the later chapters in this book assume you know how to interpret it.

8. The confusion matrix is available in the `sklearn.metrics` package. Here's how to import it and obtain the results:

```
from sklearn.metrics import confusion_matrix

confusion_matrix(df['Class'], df['Predicted'])
```

Here are the results:

```
array([[8, 0],
       [0, 6]])
```

Figure 1.8 – Classification – Evaluation with a confusion matrix

The previous figure shows that our model was able to classify every instance correctly. As a rule of thumb, if the diagonal elements stretching from the bottom left to the top right are zeros, it means the model is 100% accurate.

The confusion matrix interpretation concludes our brief refresher on supervised machine learning methods. Next, we will dive into the idea of automation, and discuss why we need it in machine learning.

Reviewing automation

This section briefly discusses the idea of automation, why we need it, and how it applies to machine learning. We will also answer the age-old question of machine learning replacing humans in their jobs, and the role of automation in that regard.

Automation plays a huge role in the modern world, and in the past centuries it has allowed us to completely remove the human factor from dangerous and repetitive jobs. This has opened a new array of possibilities on the job market, where jobs are generally based on something that cannot be automated, at least at this point in time.

But first, we have to understand what automation is.

What is automation?

There are many syntactically different definitions out there, but they all share the same basic idea. The following one presents the idea in the simplest terms:

> *Automation is a broad term that can cover many areas of technology where human input is minimized (What is Automation? – IBM, IBM Team; February 28, 2020).*

The essential part of the definition is the *minimization of the human input*. An automated process is entirely or almost entirely managed by a machine. Up to a couple of years back, machines were a great way to automate boring, routine tasks, and leave creative things to people. As you might guess, machines are not that great with creative tasks. That is, they weren't until recently.

Machine learning provides us with a mechanism to not only automate calculations, spreadsheet management, and expenses tracking, but also more cognitive tasks, such as decision making. The field evolves by the day and it's hard to say when exactly we can expect machines to take over some more creative jobs.

The concept of automation in machine learning is discussed later, but it's important to remember that machine learning can take automation to a whole other level. Not every form of automation is equal, and the generally accepted division of automation is into four levels, based on complexity:

- **Basic automation**: Automation of the simplest tasks. **Robotic Process Automation (RPA)** is the perfect example, as its goal is to use software bots to automate repetitive tasks. The end goal of this automation category is to completely remove the human factor from the equation, resulting in faster execution of repetitive tasks without error.

- **Process automation**: This uses and applies basic automation techniques to an entire business process. The end goal is to completely automate a business activity and leave humans to only give the final approval.

- **Integration automation**: This uses rules defined by humans to mimic human behavior in task completion. The end goal is to minimize human intervention in more complex business tasks.

- **AI automation**: The most complex form of automation. The goal is to have a machine that can learn and make decisions based on previous situations and the decisions made in those situations.

You now know what automation is, and next, we'll discuss why it is a must in the 21st century.

Why is automation needed?

Both companies and customers can benefit from automation. Automation can improve resource allocation and management, and can make the business scaling process easier. Due to automation, companies can provide a more reliable and consistent service, which results in a more consistent user experience. As the end result, customers are more likely to buy and spend more than if the service quality was not consistent.

In the long run, automation simplifies and reduces human activities and reduces costs. Further, any automated process is likely to perform better than the same process performed by humans. Machines don't get tired, don't have a bad day, and don't require a salary.

The following list shows some of the most important reasons for automation:

- **Time saving**: Automation simplifies daily routine tasks by making machines do them instead of humans. As the end result, humans can focus on more creative tasks right from the start.

- **Reduced cost**: Automation should be thought of as a long-term investment. It comes with some start-up costs, sure, but those are covered quickly if automation is implemented correctly.

- **Accuracy and consistency**: As mentioned before, humans are prone to errors, bad days, and inconsistencies. That's not the case with machines.

- **Workflow enhancements**: Due to automation, more time can be spent on important tasks, such as providing individual assistance to customers. Employees tend to be happier and deliver better results if their shift isn't made up solely of repetitive and routine tasks.

The difficult question is not *"do you automate?"* but rather, *"when do you automate?"* There are a lot of different opinions on this topic and there isn't a single right or wrong answer. Deciding when to automate depends on the budget you have available and on the opportunity cost (the decisions/investments you would be able to make if time was not an issue).

Automating anything you are good at and focusing on the areas that require improvement is a general rule of thumb for most companies. Even as an individual, there is a high probability that you are doing something on a daily or weekly basis that can be described in plain language. And if something can be described step by step, it can be automated.

But how does the concept of automation apply to machine learning? Are machine learning and automation synonymous? That's what we will discuss next.

Are machine learning and automation the same thing?

Well, no. But machine learning can take automation to a whole different level. Let's refer back to the four levels of automation discussed a few of paragraphs ago. Only the last one uses machine learning, and it is the most advanced form of automation.

Let's consider a single activity in our day as a *process*. If you know exactly how the process will start and end, and everything that will happen in between and in which order, then this process can be automated without machine learning.

Here's an example. For the last couple of months, you've been monitoring real-estate prices in an area you want to move to. Every morning you make yourself a cup of coffee, sit in front of a laptop, and go to a real estate website. You filter the results to see only the ads that were placed in the last 24 hours, and then enter the data, such as the location, unit price, number of rooms, and so on, into a spreadsheet.

This process takes about an hour of your day, which results in 30 hours per month. That is a lot. In 30 hours, you can easily read a book or take an online course to further develop your skills in some other area. The process described in this paragraph can be automated easily, without the need for machine learning.

Let's take a look at another example. You are spending multiple hours per day on the stock market, deciding what to buy and what to sell. This process is different from the previous one, as it involves some sort of decision making. The thing is, with all of the datasets available online, a skilled individual can use machine learning methods to automate the buy/sell decision-making process.

This is the form of automation that includes machine learning, but no, machine learning and automation are not synonymous. Each can work without the other.

The following sections discuss in great detail the role of automation in machine learning (not vice versa), and answer what we are trying to automate and how it can be achieved in the modern day and age.

Applying automation to machine learning

We've covered the idea of automation and various types of automation thus far, but what's the connection between automation and machine learning? What exactly is it that we are trying to automate in machine learning?

That's what this section aims to demystify. By the end of this section, you will know the difference between the terms *automation with machine learning* and *automating machine learning*. These two might sound similar at first, but are very different in reality.

What are we trying to automate?

Let's get one thing straight – *automation of machine learning processes* has nothing to do with *business process automation with machine learning*. In the former, we're trying to automate the machine learning itself, ergo automating the process of selecting the best model and the best hyperparameters. The latter refers to automating a business process with the help of machine learning; for example, making a decision system that decides when to buy or sell a stock based on historical data.

It's crucial to remember this distinction. The primary focus of this book is to demonstrate how automation libraries can be used to automate the process of machine learning. By doing so, you will follow the exact same approach, regardless of the dataset, and always end up with the best possible model.

Choosing an appropriate machine learning algorithm isn't an easy task. Just take a look at the following diagram:

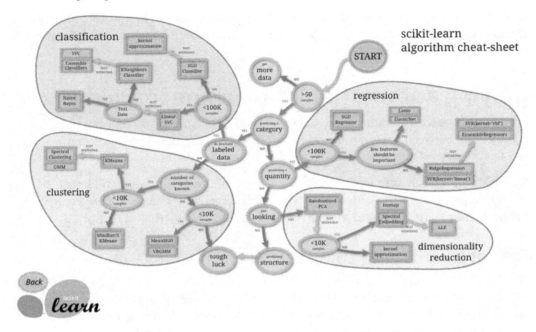

Figure 1.9 – Algorithms in scikit-learn (source: Scikit-learn: Machine Learning in Python, Pedregosa et al., JMLR 12, pp. 2825-2830, 2011)

As you can see, multiple decisions are required to select an appropriate algorithm. In addition, every algorithm has its own set of hyperparameters (parameters specified by the engineer). To make things even worse, some of these hyperparameters are continuous in nature, so when you add it all up, there are hundreds of thousands or even millions of hyperparameter combinations that you as an engineer should test for.

Every hyperparameter combination requires training and evaluation of a completely new model. Concepts such as **grid search** can help you avoid writing tens of nested loops, but it is far from an optimal solution.

Modern machine learning engineers don't spend their time and energy on model training and optimization – but instead on raising the data quality and availability. Hyperparameter tweaking can squeeze that additional 2% increase in accuracy, but it is the data *quality* that can make or break your project.

We'll dive a bit deeper into hyperparameters next and demonstrate why searching for the optimal ones manually isn't that good an idea.

The problem of too many parameters

Let's take a look at some of hyperparameters available for one of the most popular machine learning algorithms – `XGBoost`. The following list shows the general ones:

- `booster`
- `verbosity`
- `validate_parameters`
- `nthread`
- `disable_default_eval_metric`
- `num_pbuffer`
- `num_feature`

That's not much, and some of these hyperparameters are set automatically by the algorithm. The problem lies within the further selection. For example, if you choose `gbtree` as a value for the `booster` parameter, you can immediately tweak the values for the following:

- `eta`
- `gamma`
- `max_depth`
- `min_child_weight`
- `max_delta_step`
- `subsample`
- `sampling_method`
- `colsample_bytree`
- `colsample_bylevel`
- `colsample_bynode`
- `lambda`
- `alpha`
- `tree_method`
- `sketch_eps`
- `scale_pos_weight`
- `updater`

- `refresher_leaf`
- `process_type`
- `grow_policy`
- `max_leaves`
- `max_bin`
- `predictor`
- `num_parallel_tree`
- `monotone_constraints`
- `interaction_constraints`

And that's a lot! As mentioned before, some hyperparameters take in continuous values, which tremendously increases the total number of combinations. Here's the final icing on the cake – these are only hyperparameters for a single model. Different models have different hyperparameters, which makes the tuning process that much more time consuming.

Put simply, model selection and hyperparameter tuning isn't something you should do manually. There are more important tasks to spend your energy on. Even if there's nothing else you have to do, I'd prefer going for lunch instead of manual tuning any day of the week.

AutoML enables us to do just that, so we'll explore it briefly in the next section.

What is AutoML?

AutoML stands for **Automated Machine Learning**, and its primary goal is to reduce or completely eliminate the role of data scientists in building machine learning models. Hearing that sentence might be harsh at first. I know what you are thinking. But no – AutoML can't replace data scientists and other data professionals.

In the best-case scenario, AutoML technologies enable other software engineers to utilize the power of machine learning in their application, without the need to have a solid background in ML. This best-case scenario is only possible if the data is adequately gathered and prepared – a task that's not the specialty of a backend developer.

To make things even harder for the non-data scientist, the machine learning process often requires extensive feature engineering. This step can be skipped, but more often than not, this will result in poor models.

In conclusion, AutoML won't replace data scientists, rather just the contrary – it's here to make the life of data scientists easier. AutoML only automates model selection and tuning to the full extent.

There are some AutoML services that advertise themselves as fully automating even the data preparation and feature engineering jobs, but that's just by combining various features together and making something that is not interpretable most of the time. A machine doesn't know the true relationships between variables. That's the job of a data scientist to discover.

Automation options

AutoML isn't that new a concept. The idea and some implementations have been around for years, and are receiving positive feedback overall. Still, some fail to implement and fully utilize AutoML solutions in their organization due to a lack of understanding.

AutoML can't do everything – someone still has to gather the data, store it, and prepare it. This isn't a small task, and more often than not requires a significant amount of domain knowledge. Then and only then can automated solutions be utilized to their full potential.

This section explores a couple of options for implementing AutoML solutions. We'll compare one code-based tool written in Python, and one that is delivered as a browser application, meaning that no coding is required. We'll start with the code-based one first.

PyCaret

PyCaret has been widely used to make production-ready machine learning models with as little code as possible. It is a completely free solution capable of training, visualizing, and interpreting machine learning models with ease.

It has built-in support for regression and classification models and shows in an interactive way which models were used for the task, and which generated the best result. It's up to the data scientist to decide which model will be used for the task. Both training and optimization are as simple as a function call.

The library also provides an option to explain machine learning models with game-theoretic algorithms such as **SHAP** (**Shapely Additive Explanations**), also with a single function call.

PyCaret still requires a bit of human interaction. Oftentimes, though, the initialization and training process of a model must be selected explicitly by the user, and that breaks the idea of a fully-automated solution.

Further, PyCaret can be slow to run and optimize for a larger dataset. Let's take a look at a code-free AutoML solution next.

ObviouslyAI

Not all of us know how to develop machine learning models, or even how to write code. That's where drag and drop solutions come into play. **ObviouslyAI** is certainly one of the best out there, and is also easy to use.

This service allows for in-browser model training and evaluation, and can even explain the reasoning behind decisions made by a model. It's a no-brainer for companies in which machine learning isn't the core business, as it's pretty easy to start with and doesn't cost nearly as much as an entire data science team.

A big gotcha with services like this one is the pricing. There's always a free plan included, but in this particular case it's limited to datasets with fewer than 50,000 rows. That's completely fine for occasional tests here and there, but is a deal-breaker for most production use cases.

The second deal-breaker is the actual automation. You can't easily automate mouse clicks and dataset loads. This service automates the machine learning process itself completely, but you still have to do some manual work.

TPOT

The acronym **TPOT** stands for **Tree-based Pipeline Optimization Tool**. It is a Python library designed to handle machine learning tasks in an automated fashion.

Here's a statement from the official documentation:

> *Consider TPOT your Data Science Assistant. TPOT is a Python Automated Machine Learning tool that optimizes machine learning pipelines using genetic programming (TPOT Documentation page, TPOT Team; November 5, 2019).*

Genetic programming is a term that is further discussed in the later chapters. For now, just know that it is based on **evolutionary algorithms** – a special type of algorithm used to discover solutions to problems that humans don't know how to solve.

In a way, TPOT *is* your data science assistant. You can use it to automate everything boring in a data science project. The term "boring" is subjective, but throughout the book, we use it to refer to the tasks of manually selecting and tweaking machine learning models (read: spending days waiting for the model to tune).

TPOT can't automate the process of data gathering and cleaning, and the reason is obvious – a machine can't read your mind. It can, however, perform machine learning tasks on well prepared datasets better than most data scientists.

The following chapters discuss the library in great detail.

Summary

You've learned a lot in this section – or had a brief recap, at least. You are now fresh on the concepts of machine learning, regression, classification, and automation. All of these are crucial for the following, more demanding sections.

The chapters after the next one will dive deep into the code, so you will get a full grasp of the library. Everything from the most basic regression and classification automation, to parallel training, neural networks, and model deployment will be discussed.

In the next chapter, we'll dive deep into the TPOT library, its use cases, and its underlying architecture. We will discuss the core principle behind TPOT – genetic programming – and how is it used to solve regression and classification tasks. We will also fully configure the environment for the Windows, macOS, and Linux operating systems.

Q&A

1. In your own words, define the term "machine learning."

2. Explain supervised learning in a couple of sentences.

3. What's the difference between regression and classification machine learning tasks?

4. Name three areas where machine learning is used and provide concrete examples.

5. How would you describe automation?

6. Why do we need automation in this day and age?

7. What's the difference between terms "automation with machine learning" and "machine learning automation"?

8. Are the terms "machine learning" and "automation" synonyms? Explain your answer.

9. Explain the problem of too many parameters in manual machine learning.

10. Define and briefly explain AutoML.

Further reading

Here are the sources we referenced in this chapter:

- *A Brief History of Machine Learning*: `https://www.dataversity.net/a-brief-history-of-machine-learning/`

- *What is Machine Learning? A definition*: `https://expertsystem.com/machine-learning-definition/`

- *Machine Learning*: `https://www.investopedia.com/terms/m/machine-learning.asp`

- *Machine Learning Definition*: `https://deepai.org/machine-learning-glossary-and-terms/machine-learning`

- *What is automation?*: `https://www.ibm.com/topics/automation`

- *TPOT*: `https://github.com/EpistasisLab/tpot`

Section 2: TPOT – Practical Classification and Regression

This section provides a deep dive into what TPOT is, what it isn't, how it is used to handle automation in machine learning, and what types of tasks TPOT can automate. This section also shows how to install TPOT and the environment setup in general.

This section contains the following chapters:

2
Deep Dive into TPOT

In this chapter, you'll learn everything about the theoretical aspects of the TPOT library and its underlying architecture. Topics such as architecture and **genetic programming (GP)** will be crucial to having a full grasp of the inner workings of the library.

We will go through TPOT use cases and dive deep into different approaches to solve various machine learning problems. You can expect to learn the basics of automation in regression and classification tasks.

We will also go through a complete environment setup for standalone Python installation and the Anaconda distribution and show you how to set up a virtual environment.

This chapter will cover the following topics:

- Introducing TPOT
- Types of problems TPOT can solve
- Installing TPOT and setting up the environment

Technical requirements

To complete this chapter, you only need a computer with Python installed. Both the standalone version and Anaconda are fine. We'll go through the installation for both through a virtual environment toward the end of the chapter.

There is no code for this chapter.

Introducing TPOT

TPOT, or **Tree-based Pipeline Optimization Tool**, is an open source library for performing machine learning in an automated fashion with the Python programming language. Below the surface, it uses the well-known **scikit-learn** machine learning library to perform data preparation, transformation, and machine learning. It also uses GP procedures to discover the best-performing pipeline for a given dataset. The concept of GP is covered in later sections.

As a rule of thumb, you should use TPOT every time you need an automated machine learning pipeline. Data science is a broad field, and libraries such as TPOT enable you to spend much more time on data gathering and cleaning, as everything else is done automatically.

The following figure shows what a typical machine learning pipeline looks like:

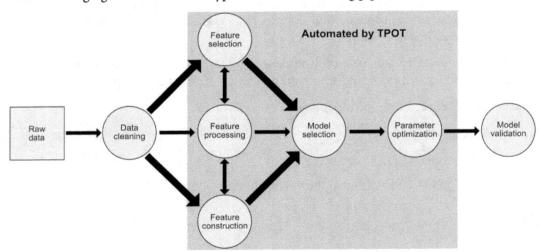

Figure 2.1 – Example machine learning pipeline

The preceding figure shows which parts of a machine learning process can and can't be automated. The data gathering phase (**Raw data**) is essential for any machine learning project. In this phase, you gather data that will serve as input to a machine learning model. If the input data isn't good enough, or there's not enough of it, machine learning algorithms can't produce good-quality models.

Assuming there's enough data and you can access it, the next most significant problem is data cleaning. This step can't be automated, at least not entirely, for obvious reasons. Every dataset is different; hence there's no single approach to data cleaning. Missing and misformatted values are the most common and the most time-consuming types of problem, and they typically require a substantial amount of domain knowledge to address successfully.

Once you have a fair amount of well-prepared data, TPOT can come into play. TPOT uses GP to find the best algorithm for a particular task, so there's no need to manually choose and optimize a single algorithm. The *Darwinian process of natural selection* inspires genetic algorithms, but more on that in a couple of sections.

The TPOT pipeline has many parameters, depending on the type of problem you are trying to solve (regression or classification). All of the parameters are discussed later in the chapter, but these are the ones you should know regardless of the problem type:

- `generations`: Represents the number of iterations the pipeline optimization process is run for
- `population_size`: Represents the number of individuals to retain in the GP population in every generation
- `offspring_size`: Represents the number of offspring to produce in each generation
- `mutation_rate`: Tells the GP algorithm how many pipelines to apply random changes to every generation
- `crossover_rate`: Tells the GP algorithm how many pipelines to breed every generation
- `cv`: Cross-validation technique used for evaluating pipelines
- `scoring`: A function that is used to evaluate the quality of a given pipeline

Once TPOT finishes with the optimization, it returns Python code for the best pipeline it found so you can proceed with model evaluation and validation on your own. A simplified example of a TPOT pipeline is shown in the following figure:

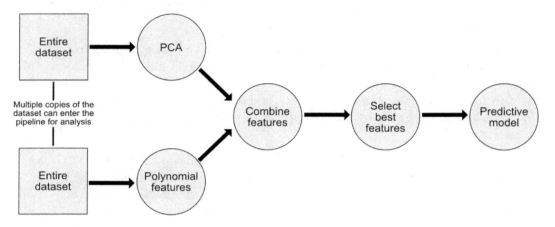

Figure 2.2 – Example TPOT pipeline

The TPOT library was built on top of Python's well-known machine learning package scikit-learn. As a result, TPOT has access to all of its classes and methods. The preceding figure shows **PCA** and **Polynomial features** as two possible feature preprocessing operations. TPOT isn't limited to these two, but instead can use any of the following:

- PCA
- RandomizedPCA
- PolynomialFeatures
- Binarizer
- StandardScaler
- MinMaxScaler
- MaxAbsScaler
- RobustScaler

These are all classes built into scikit-learn, used to modify the dataset in some way and return a modified dataset. The next step involves some kind of feature selection. This step aims to select only the features with good predictive power and discard the others. By doing so, TPOT is reducing the dimensionality of the machine learning problem, which as an end result makes the problem easier to solve.

The preceding figure hides this abstraction behind the **Select best features** step. To perform this step, TPOT can use one of the following algorithms:

- `VarianceThreshold`
- `SelectKBest`
- `SelectPercentile`
- `SelectFwe`
- `RecursiveFeatureElimination`

As you can see, TPOT is very flexible when it comes to model training approaches. To understand further what's going on below the surface, we'll need to cover a bit of GP. The following section does that.

A brief overview of genetic programming

GP is a type of evolutionary algorithm, a subset of machine learning (*Genetic Programming page, GP Team; June 1, 2019*). Evolutionary algorithms are used for finding solutions to problems that we as humans don't know how to solve directly. These algorithms generate solutions that are, at worst, comparable to the best human solutions, and often better.

In machine learning, GP can be used to discover the relationship between features in a dataset (**regression**), and to group data into categories (**classification**). In regular software engineering, GP is applied through code synthesis, genetic improvement, automatic bug fixing, and in developing game-playing strategies (*Genetic Programming page, GP Team; June 1, 2019*).

GP is inspired by biological evolution and its mechanisms. It uses algorithms based on random mutation, crossover, fitness functions, and generations to solve the previously described regression and classification tasks for machine learning. These properties should sound familiar, as we covered them in the previous section.

The idea behind GP is essential for advancements in machine learning because it is based on the *Darwinian process of natural selection*. In machine learning terms, these processes are used to generate optimal solutions – models and hyperparameters.

Genetic algorithms have three properties:

- **Selection**: Consists of a population of possible solutions and the fitness function. Each fit is evaluated at every iteration.

- **Crossover**: The process of selecting the best (fittest) solution and performing crossover to create a new population.
- **Mutation**: Taking the children from the previous point and mutating them with some random modifications until the best solution is obtained.

It is always a good idea to know the basics and the underlying architecture of the language/library you are dealing with. TPOT is user-friendly and easy to use, so it doesn't require us to know everything about GP and genetic algorithms. For that reason, this chapter won't dive deeper into the topic. If you are interested in learning more about GP, you'll find useful links at the end of the chapter.

We've discussed a lot about the good sides of machine learning automation, TPOT, and GP. But are there any downsides? The following section addresses a couple of them.

TPOT limitations

Thus far, we have discussed only the good things about the TPOT library and the automation of machine learning processes in general. In this case, the pros outweigh the cons, but we should still talk about potential downsides. The first one is the execution time. It will vary based on the size of the dataset and the hardware specifications, but in general, it will take a lot of time to finish – hours or days for large datasets and minutes for smaller ones.

It is essential to understand why. With the default TPOT settings – 100 generations with 100 population sizes – TPOT will evaluate 10,000 pipelines before finishing. That is equivalent to performing feature engineering and training of a machine learning model 10,000 times. Because of this, TPOT is expected to run slowly.

Things get more complicated if you decide to bring **cross-validation** into the picture. This term represents a procedure where a machine learning model is trained k times on $k - 1$ subsets and evaluated on a separate subset. The goal of cross-validation is to have a more accurate representation of the model's performance. The choice of k is arbitrary, but in practice, the most common value is 10.

In practice, cross-validation makes TPOT significantly slower. When using cross-validation, TPOT will evaluate 100 generations with 100 population sizes and 10 cross-validation folds by default. This results in 100,000 different pipelines to evaluate before finishing.

To address this issue, TPOT introduced the `max_time_mins` parameter. It is set to `None` by default, but you can set its value explicitly to any integer. For example, specifying `max_time_mins=10` would give TPOT only 10 minutes to optimize the pipeline. It's not an ideal solution if you want the best results, but it comes in handy when you are on a tight schedule.

The second downside is that TPOT can sometimes recommend different solutions (pipelines) for the same dataset. This will often be a problem when the TPOT optimizer is run for a short amount of time. For example, if you have used the `max_time_mins` parameter to limit how long the optimizer would run, it's not a surprise that you will get a slightly different "optimal" pipeline every time.

This isn't a reason to worry, as all pipelines should still outperform anything you can do manually in the same time frame, but it is essential to know why this happens. There are two possible reasons:

- *The TPOT optimizer didn't converge*: This is the most likely scenario. TPOT wasn't able to find an optimal pipeline due to lack of time, or the dataset was too complex to optimize for in the given time period (or both).

- *There are multiple "optimal" pipelines*: It's not uncommon to see numerous approaches working identically for some machine learning problems. This is a more likely scenario if the dataset is relatively small.

This section briefly introduced the TPOT library and explained its benefits and shortcomings. The next section goes over the types of problems TPOT is solving and discusses the automation of regression and classification tasks in great detail.

Types of problems TPOT can solve

The TPOT library was designed as a go-to tool for automating machine learning tasks; hence, it should be able to handle with ease anything you throw at it. We will start using TPOT in a practical sense soon. *Chapter 3, Exploring before Regression*, shows how to use the library to handle practical tasks with many examples, and the following chapters focus on other types of tasks.

In general, TPOT can be used to handle the following types of tasks:

- **Regression**: Where the target variable is continuous, such as age, height, weight, score, or price. Refer to *Chapter 1, Machine Learning and the Idea of Automation*, for a brief overview of regression.

- **Classification**: Where the target variable is categorical, such as sold/did not sell, churn/did not churn, or yes/no. Refer to *Chapter 1, Machine Learning and the Idea of Automation*, for a brief overview of classification.

- **Parallel training**: TPOT can handle the training of machine learning models in a parallel manner through the **Dask** library. Please read *Chapter 5, Parallel Training with TPOT and Dask*, to get a full picture.

- **Neural networks**: TPOT can even build models based on state-of-the-art neural network algorithms in a fully automated fashion. Please read *Chapter 6, Getting Started with Deep Learning – Crash Course in Neural Networks*, for a quick crash course on neural networks, and *Chapter 7, Neural Network Classifier with TPOT*, for the practical implementation of TPOT.

The rest of this section briefly discusses how TPOT handles regression and classification tasks and spends a good amount of time exploring and explaining their parameters, attributes, and functions. You will learn how TPOT handles parallel training with Dask in *Chapter 5, Parallel Training with TPOT and Dask*, and how it handles neural networks in *Chapter 6, Getting Started with Deep Learning – Crash Course in Neural Networks*, because these topics require covering the prerequisites first.

How TPOT handles regression tasks

The TPOT library handles regression tasks through the `tpot.TPOTRegressor` class. This class performs an intelligent search over machine learning pipelines containing supervised regression models, preprocessors, feature selection techniques, and any other estimator or transformer that follows the `scikit-learn` API (*TPOT Documentation page, TPOT Team; November 5, 2019*).

The same class also performs a search over the hyperparameters of all objects in a pipeline. The `tpot.TPOTRegressor` class allows us to fully customize the models, transformers, and parameters searched over through the `config_dict` parameter.

We will now go over the parameters that the `tpot.TPOTRegressor` class expects when instantiated:

- `generations`: Integer or None (default = `100`). An optional parameter that specifies the number of iterations to run the pipeline optimization process. It must be positive. If it is not defined, the `max_time_mins` parameter must be specified instead.

- `population_size`: Integer (default = `100`). An optional parameter that specifies the number of individuals to retain in the GP population in every generation. Must be a positive number.

- `offspring_size`: Integer (default = the same as `population_size`). An optional parameter used to specify the number of offsprings to produce in each GP generation. Must be a positive number.

- `mutation_rate`: Float (default = `0.9`). An optional parameter used to specify the mutation rate for the GP algorithm. Must be in the range [0.0, 1.0]. This parameter is used to instruct the algorithm on how many pipelines to apply random changes to every generation.

- `crossover_rate`: Float (default = `0.1`). An optional parameter that instructs the GP algorithm on how many pipelines to "breed" every generation. Must be in the range [0.0, 1.0].

- `scoring`: String or callable (default = `neg_mean_squared_error`). An optional parameter used to specify the function name for regression pipeline evaluation. Can be `neg_median_abs_value`, `neg_mean_abs_error`, `neg_mean_squared_error`, or `r2`.

- `cv`: Integer, cross-validation generator, or an iterable (default = `5`). An optional parameter used to specify a cross-validation strategy for evaluating regression pipelines. If the passed value is an integer, it expects the number of folds. In other cases, it expects an object to be used as a cross-validation generator, or an iterable yielding train/test splits, respectively.

- `subsample`: Float (default = `1.0`). An optional parameter used to specify a value for a fraction of training samples used in the optimization process. Must be in the range [0.0, 1.0].

- `n_jobs`: Integer (default = `1`). An optional parameter used to specify the number of processes to use in parallel for the evaluation of pipelines during optimization. Set it to `-1` to use all CPU cores. Set it to `-2` to use all but one CPU core.

- `max_time_mins`: Integer or None (default = `None`). An optional parameter used to specify how many minutes TPOT can perform the optimization. TPOT will optimize for less time only if all of the generations evaluate before the specified max time in minutes.

- `max_eval_time_mins`: Float (default = `5`). An optional parameter used to specify how many minutes TPOT has to evaluate a single pipeline. If the parameter is set to a high enough value, TPOT will evaluate more complex pipelines. At the same time, it also makes TPOT run longer.

- `random_state`: Integer or None (default = None). An optional parameter used to specify the seed for a pseudo-random number generator. Use it to get reproducible results.

- `config_dict`: Dictionary, string, or None (default = None). An optional parameter used to specify a configuration dictionary for customizing the operators and parameters that TPOT searches during optimization. Possible inputs are as follows:

  a) *None*: TPOT uses the default configuration.

  b) *Python dictionary*: TPOT uses your configuration.

  c) *'TPOT light'*: String; TPOT will use a built-in configuration with only fast models and preprocessors.

  d) *'TPOT MDR'*: String; TPOT will use a built-in configuration specialized for genomic studies.

  e) *'TPOT sparse'*: String; TPOT will use a configuration dictionary with a one-hot encoder and operators that support sparse matrices.

- `template`: String (default = None). An optional parameter used to specify a template of a predefined pipeline. Used to specify the desired structure for the machine learning pipeline evaluated by TPOT.

- `warm_start`: Boolean (default = False). An optional parameter used to indicate whether the current instance should reuse the population from previous calls to the `fit()` function. This function is discussed later in the chapter.

- `memory`: A memory object or a string (default = None). An optional parameter used to cache each transformer after calling the `fit()` function. This function is discussed later in the chapter.

- `use_dask`: Boolean (default = False). An optional parameter used to specify whether *Dask-ML's* pipeline optimizations should be used.

- `periodic_checkpoint_folder`: Path string (default = None). An optional parameter used to specify in which folder TPOT will save pipelines while optimizing.

- `early_stop`: Integer (default = None). An optional parameter used to specify after how many generations TPOT will stop optimizing if there's no improvement.

- verbosity: Integer (default = 0). An optional parameter used to specify how much information TPOT outputs to the console while running. Possible options are as follows:

  a) *0*: TPOT doesn't print anything.

  b) *1*: TPOT prints minimal information.

  c) *2*: TPOT prints more information and provides a progress bar.

  d) *3*: TPOT prints everything and provides a progress bar.

- disable_update_check: Boolean (default = False). An optional parameter that indicates whether the TPOT version checker should be disabled. You can ignore this parameter because it only tells you whether a newer version of the library is available, and has nothing to do with the actual training.

That's a lot of parameters you should know about if your goal is to truly master the library – at least the part of it that handles regression tasks. We've only covered parameters for the tpot.TPOTRegressor class and we will discuss attributes and functions next. Don't worry; there are only a couple of them available.

Let's start with attributes. There are three in total. These become available once the pipeline is fitted:

- fitted_pipeline_: Pipeline object from scikit-learn. Shows you the best pipeline that TPOT discovered during the optimization for a given training set.

- pareto_front_fitted_pipelines_: Python dictionary. It contains all the pipelines on the TPOT Pareto front. The dictionary key is the string representing the pipeline, and the value is the corresponding pipeline. This argument is available only when the verbosity parameter is set to 3.

- evaluated_individuals_: Python dictionary. It contains all evaluated pipelines. The dictionary key is the string representing the pipeline, and the value is a tuple containing the number of steps in each pipeline and the corresponding accuracy metric.

We will see how the mentioned attributes work in practice in the following chapters. The only thing left to discuss for this section are functions belonging to the `tpot.` `TPOTRegressor` class. There are four in total:

- `fit(features, target, sample_weight=None, groups=None)`: This function is used to run the TPOT optimization process. The `features` parameter is an array of the features/predictors/attributes used for predicting the target variable. The `target` parameter is also an array that specifies the list of target labels for prediction. The other two parameters are optional. The `sample_weights` parameter is an array indicating per-sample weights. Higher weights indicate more importance. The last parameter, `groups`, is an array that specifies group labels for the samples used when performing cross-validation. It should only be used in conjunction with group cross-validation functions. The `fit()` function returns a copy of the fitted TPOT object.

- `predict(features)`: This function is used to generate new predictions based on the `features` parameter. This parameter is an array containing features/predictors/attributes for predicting the target variable. The function returns an array of predictions.

- `score(testing_features, testing_target)`: This function returns a score of the optimized pipeline on a given testing data. The function accepts two parameters. The first one is `testing_features`. It is an array/feature matrix of the testing set. The second one is `testing_target`. It is also an array, but of target labels for prediction in the training set. The function returns the accuracy score on the test set.

- `export(output_file_name)`: This function is used to export the optimized pipeline as Python code. The function accepts a single parameter, `output_file_name`. It is used to specify the path and a filename where the Python code should be stored. If the value for the mentioned parameter isn't specified, the whole pipeline is returned as text.

With this overview of parameters, attributes, and functions, you are ready to use TPOT's regression capabilities in practice. *Chapter 3, Exploring before Regression*, is packed with regression examples, so don't hesitate to jump to it if you want to automate regression tasks.

The next section of this chapter discusses how TPOT handles classification tasks.

How TPOT handles classification tasks

The TPOT library handles classification tasks through the `tpot.TPOTClassifer` class. This class performs a search over machine learning pipelines containing supervised regression models, preprocessors, feature selection techniques, and any other estimator or transformer that follows the `scikit-learn` API (*TPOT Documentation page, TPOT Team; November 5, 2019*). The class also performs a search over the hyperparameters of all objects in the pipeline.

The `tpot.TPOTClassifier` class allows us to fully customize the models, transformers, and parameters that will be searched over through the `config_dict` parameter.

The `tpot.TPOTClassifier` class contains mostly the same parameters, attributes, and functions that the previously discussed `tpot.TPOTRegressor` has, so going over all of them again in detail would be redundant. Instead, we will just mention the identical parameters, attributes, and functions, and we will introduce and explain the ones that are unique for classification or work differently.

First, let's go over the parameters:

- `generations`
- `population_size`
- `offspring_size`
- `mutation_rate`
- `crossover_rate`
- `scoring`: String or callable (default = `accuracy`). This is an optional parameter used to evaluate the quality of a given pipeline for the classification problem. The following scoring functions can be used: `accuracy`, `adjusted_rand_score`, `average_precision`, `balanced_accuracy`, `f1`, `f1_macro`, `f1_micro`, `f1_samples`, `f1_weighted`, `neg_log_loss`, `precision`, `recall`, `recall_macro`, `recall_micro`, `recall_samples`, `recall_weighted`, `jaccard`, `jaccard_macro`, `jaccard_micro`, `jaccard_samples`, `jaccard_weighted`, `roc_auc`, `roc_auc_ovr`, `roc_auc_ovo`, `roc_auc_ovr_weighted`, or `roc_auc_ovo_weighted`. If you want to use a custom scoring function, you can pass it as a function with the following signature: `scorer(estimator, X, y)`.
- `cv`
- `subsample`
- `n_jobs`

- `max_time_mins`
- `max_eval_time_mins`
- `random_state`
- `config_dict`
- `template`
- `warm_start`
- `memory`
- `use_dask`
- `periodic_checkpoint_folder`
- `early_stop`
- `verbosity`
- `disable_update_check`
- `log_file`: File-like class or string (default = None). This is an optional parameter used to save progress content to a file. If a string value is provided, it should be the path and the filename of the desired output file.

As we can see, one parameter has changed, and one parameter is completely new. To repeat – please refer to the preceding subsection for detailed clarifications on what every parameter does.

Next, we have to discuss the attributes of the `tpot.TPOTClassifier` class. These become available once the pipeline optimization process is finished. There are three in total, and all of them behave identically to the `tpot.TPOTRegressor` class:

- `fitted_pipeline_`
- `pareto_front_fitted_pipelines_`
- `evaluated_individuals_`

Finally, we will discuss functions. As with the parameters, all are listed, but only the new and changed ones are discussed in detail. There are five functions in total:

- `fit(features, classes, sample_weight=None, groups=None)`: This function behaves identically to the one in `tpot.TPOTRegressor`, but the second parameter is called `classes` instead of `target`. This parameter expects an array of class labels for prediction.

- `predict(features)`
- `predict_proba(features)`: This function does the same task as the `predict()` function but returns class probabilities instead of classes. You can see where the model was completely certain about predictions and where it wasn't so certain by examining the probabilities. You can also use class probabilities to adjust the decision threshold. You will learn how to do that in *Chapter 4, Exploring before Classification*.
- `score(testing_features, testing_target)`
- `export(output_file_name)`

You are now ready to see how TPOT works in practice. Most of the time, there's no need to mess around with some of the listed parameters, but you need to know that they exist for more advanced use cases. *Chapter 4, Exploring before Classification*, is packed with classification examples, so don't hesitate to jump to it if you want to learn how to automate classification tasks.

The next section of this chapter discusses how to set up a TPOT environment through a virtual environment, both for standalone Python installation and installation through Anaconda.

Installing TPOT and setting up the environment

This section discusses the last required step before diving into the practical stuff – installation and environment setup. It is assumed that you have Python 3 installed, either through the standalone installation or through Anaconda.

You will learn how to set up a virtual environment for TPOT for the following scenarios:

- Standalone Python
- Anaconda

There's no need to read both installation sections, so just pick whichever suits you better. There shouldn't be any difference with regards to installation between operating systems. If you have Python installed as a standalone installation, you have access to `pip` through the terminal. If you have it installed through Anaconda, you have access to Anaconda Navigator.

Installing and configuring TPOT with standalone Python installation

Before proceeding, make sure to have Python and `pip` (package manager for Python) installed. You can check whether `pip` is installed by entering the following line of code into the terminal:

```
> pip
```

If you see output like the one in the following figure, you are good to go:

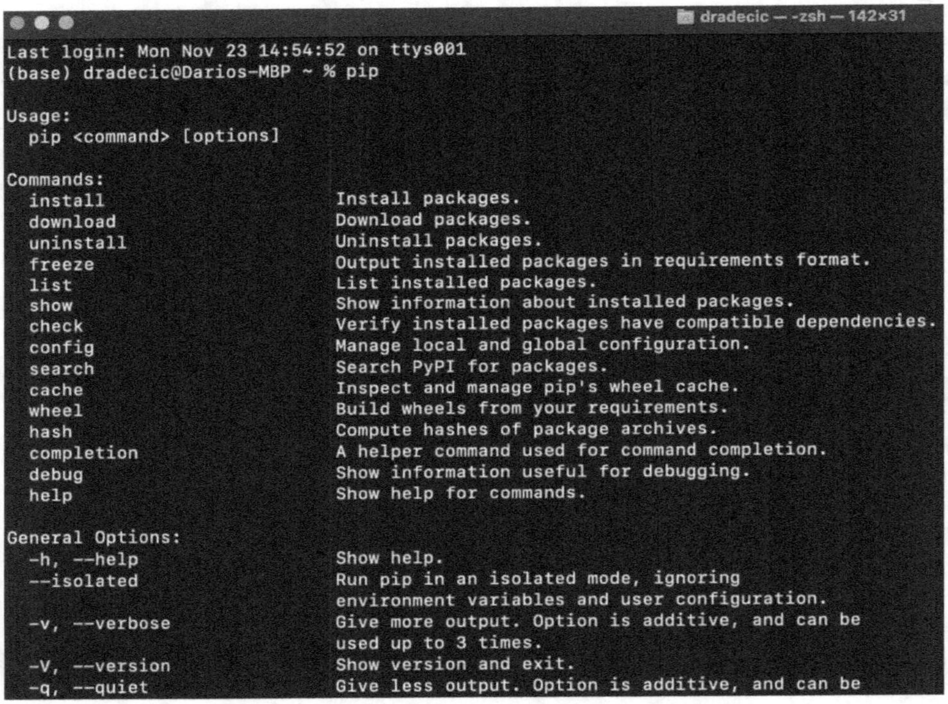

Figure 2.3 – Checking whether pip is installed

We can now proceed to the virtual environment setup:

1. The first thing to do is to install the `virtualenv` package. To do so, execute this line from the terminal:

    ```
    > pip install virtualenv
    ```

2. After a couple of seconds, you should see a success message, as shown in the following figure:

```
(base) dradecic@Darios-MBP ~ % pip install pipenv
Collecting pipenv
  Downloading pipenv-2020.11.15-py2.py3-none-any.whl (3.9 MB)
     |                              | 3.9 MB 4.7 MB/s
Collecting virtualenv-clone>=0.2.5
  Downloading virtualenv_clone-0.5.4-py2.py3-none-any.whl (6.6 kB)
Collecting virtualenv
  Downloading virtualenv-20.2.1-py2.py3-none-any.whl (4.9 MB)
     |                              | 4.9 MB 13.0 MB/s
Requirement already satisfied: setuptools>=36.2.1 in ./opt/anaconda3/lib/python3.8/site-packages (from pipenv) (49.2.0.post20200714)
Requirement already satisfied: certifi in ./opt/anaconda3/lib/python3.8/site-packages (from pipenv) (2020.6.20)
Requirement already satisfied: pip>=18.0 in ./opt/anaconda3/lib/python3.8/site-packages (from pipenv) (20.1.1)
Requirement already satisfied: six<2,>=1.9.0 in ./opt/anaconda3/lib/python3.8/site-packages (from virtualenv->pipenv) (1.15.0)
Collecting distlib<1,>=0.3.1
  Downloading distlib-0.3.1-py2.py3-none-any.whl (335 kB)
     |                              | 335 kB 19.4 MB/s
Requirement already satisfied: filelock<4,>=3.0.0 in ./opt/anaconda3/lib/python3.8/site-packages (from virtualenv->pipenv) (3.0.12)
Collecting appdirs<2,>=1.4.3
  Downloading appdirs-1.4.4-py2.py3-none-any.whl (9.6 kB)
Installing collected packages: virtualenv-clone, distlib, appdirs, virtualenv, pipenv
Successfully installed appdirs-1.4.4 distlib-0.3.1 pipenv-2020.11.15 virtualenv-20.2.1 virtualenv-clone-0.5.4
```

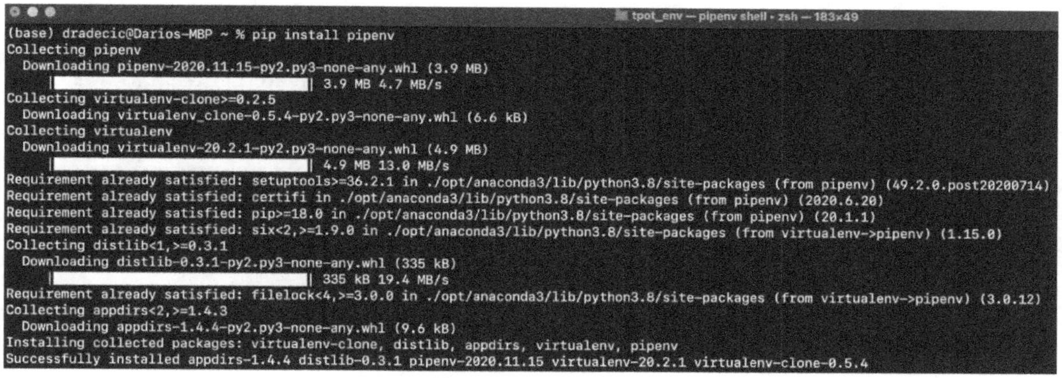

Figure 2.4 – virtualenv installation

3. The next step is to create a folder where the TPOT environment will be stored. Ours is located in the Documents folder, but you can store it anywhere. Here are the exact shell lines you need to execute to create the folder and install the Python virtual environment:

```
> cd Documents
> mkdir python_venvs
> virtualenv python_venvs/tpot_env
```

4. The execution and results are shown in the following figure:

```
(base) dradecic@Darios-MBP ~ % cd Documents
(base) dradecic@Darios-MBP Documents % mkdir tpot_env
(base) dradecic@Darios-MBP Documents % cd tpot_env
(base) dradecic@Darios-MBP tpot_env % pipenv install --python 3.9
Warning: the environment variable LANG is not set!
We recommend setting this in ~/.profile (or equivalent) for proper expected behavior.
Creating a virtualenv for this project...
Pipfile: /Users/dradecic/Documents/tpot_env/Pipfile
Using /usr/local/bin/python3.9 (3.9.0) to create virtualenv...
⠋ Creating virtual environment...created virtual environment CPython3.9.0.final.0-64 in 1023ms
  creator CPython3Posix(dest=/Users/dradecic/.local/share/virtualenvs/tpot_env-Vd9ZHFbz, clear=False, no_vcs_ignore=False, global=False)
  seeder FromAppData(download=False, pip=bundle, setuptools=bundle, wheel=bundle, via=copy, app_data_dir=/Users/dradecic/Library/Application Support/virtualenv)
    added seed packages: pip==20.2.4, setuptools==50.3.2, wheel==0.35.1
  activators BashActivator,CShellActivator,FishActivator,PowerShellActivator,PythonActivator,XonshActivator

✔ Successfully created virtual environment!
Virtualenv location: /Users/dradecic/.local/share/virtualenvs/tpot_env-Vd9ZHFbz
Creating a Pipfile for this project...
Pipfile.lock not found, creating...
Locking [dev-packages] dependencies...
Locking [packages] dependencies...
Updated Pipfile.lock (16c839)!
Installing dependencies from Pipfile.lock (16c839)...
  🐍   ▮▮▮▮▮▮▮▮▮▮▮▮▮▮▮▮▮▮▮▮▮▮ 0/0 —
To activate this project's virtualenv, run pipenv shell.
Alternatively, run a command inside the virtualenv with pipenv run.
```

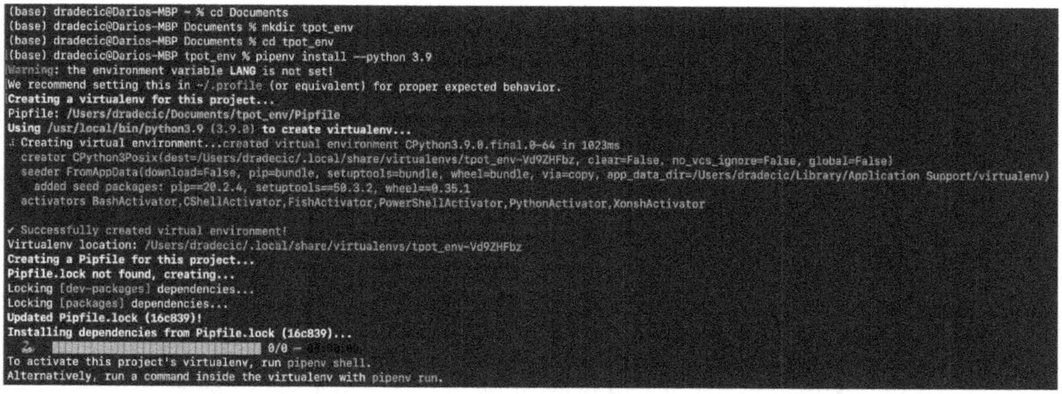

Figure 2.5 – Creating a virtual environment

The environment is successfully installed now.

5. To activate the environment, you need to execute the following line from the terminal:

```
> source python_venvs/tpot_env/bin/activate
```

6. The text in parentheses confirms that the environment is activated. Take a look at the change from the `base` environment to `tpot_env` in the following figure:

```
(base) dradecic@Darios-MBP tpot_env % pipenv shell
Launching subshell in virtual environment...
 . /Users/dradecic/.local/share/virtualenvs/tpot_env-Vd9ZHFbz/bin/activate
(base) dradecic@Darios-MBP tpot_env %  . /Users/dradecic/.local/share/virtualenvs/tpot_env-Vd9ZHFbz/bin/activate
(tpot_env) (base) dradecic@Darios-MBP tpot_env %
```

Figure 2.6 – Activating a virtual environment

7. To deactivate the environment, enter the following line into the terminal:

```
> deactivate
```

You can see the results in the following figure:

```
(tpot_env) (base) dradecic@Darios-MBP tpot_env % deactivate
(base) dradecic@Darios-MBP tpot_env %
```

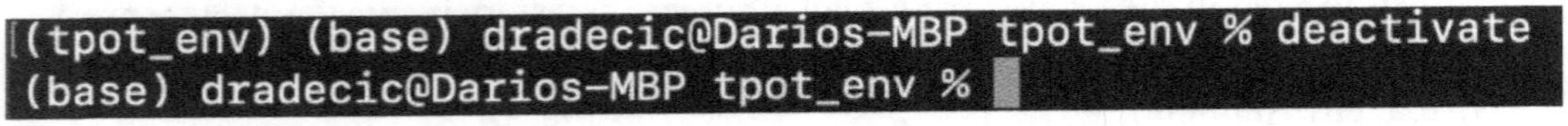

Figure 2.7 – Deactivating a virtual environment

We now have everything needed to begin with the library installation. Throughout the entire book, we will need the following:

- `jupyterlab`: A notebook environment required for analyzing and exploring data and building machine learning models in an interactive way.

- `numpy`: Python's go-to library for numerical computations.

- `pandas`: A well-known library for data loading, processing, preparation, transformation, aggregation, and even visualization.

- `matplotlib`: Python's standard data visualization library. We will use it sometimes for basic plots.

- `seaborn`: A data visualization library with more aesthetically pleasing visuals than `matplotlib`.

- `scikit-learn`: Python's go-to library for machine learning and everything related to it.

- TPOT: Used to find optimal machine learning pipelines in an automated fashion.

8. To install every mentioned library, you can execute the following line from the opened terminal window:

```
> pip install jupyterlab numpy pandas matplotlib seaborn
  scikit-learn TPOT
```

Python will immediately start downloading and installing libraries, as shown in the following figure:

```
(tpot_env) (base) dradecic@Darios-MBP tpot_env % pip install jupyterlab numpy pandas matplotlib seaborn scikit-learn TPOT
Collecting jupyterlab
  Using cached jupyterlab-2.2.9-py3-none-any.whl (7.9 MB)
Collecting numpy
  Using cached numpy-1.19.4-cp38-cp38-macosx_10_9_x86_64.whl (15.3 MB)
Collecting pandas
  Using cached pandas-1.1.4-cp38-cp38-macosx_10_9_x86_64.whl (10.1 MB)
Collecting matplotlib
  Using cached matplotlib-3.3.3-cp38-cp38-macosx_10_9_x86_64.whl (8.5 MB)
Collecting seaborn
  Using cached seaborn-0.11.0-py3-none-any.whl (283 kB)
Collecting scikit-learn
  Using cached scikit_learn-0.23.2-cp38-cp38-macosx_10_9_x86_64.whl (7.2 MB)
Collecting TPOT
  Using cached TPOT-0.11.6.post1-py3-none-any.whl (86 kB)
Collecting jinja2>=2.10
  Using cached Jinja2-2.11.2-py2.py3-none-any.whl (125 kB)
Collecting jupyterlab-server<2.0,>=1.1.5
  Using cached jupyterlab_server-1.2.0-py3-none-any.whl (29 kB)
Collecting notebook>=4.3.1
  Using cached notebook-6.1.5-py3-none-any.whl (9.5 MB)
Collecting tornado!=6.0.0,!=6.0.1,!=6.0.2
  Using cached tornado-6.1-cp38-cp38-macosx_10_9_x86_64.whl (416 kB)
Collecting python-dateutil>=2.7.3
  Using cached python_dateutil-2.8.1-py2.py3-none-any.whl (227 kB)
Collecting pytz>=2017.2
```

Figure 2.8 – Installing libraries with pip

9. To test whether the environment was successfully configured, we can open JupyterLab from the terminal. Execute the following shell command once the libraries are installed:

```
> jupyter lab
```

If you see something similar to the following, then everything went according to plan. The browser window with Jupyter should open immediately:

```
(tpot_env) dradecic@Darios-MBP ~ % jupyter lab
[I 14:56:31.709 LabApp] The port 8888 is already in use, trying another port.
[I 14:56:31.710 LabApp] The port 8889 is already in use, trying another port.
[I 14:56:31.946 LabApp] JupyterLab extension loaded from /Users/dradecic/opt/anaconda3/envs/tpot_env/lib/python3.8/site-packages/jupyterlab
[I 14:56:31.947 LabApp] JupyterLab application directory is /Users/dradecic/opt/anaconda3/envs/tpot_env/share/jupyter/lab
[I 14:56:31.949 LabApp] Serving notebooks from local directory: /Users/dradecic/Library/Mobile Documents/com~apple~CloudDocs
[I 14:56:31.949 LabApp] Jupyter Notebook 6.1.5 is running at:
[I 14:56:31.949 LabApp] http://localhost:8890/?token=d739f9a1a3539248da41cd364ba8b5eae380fca905393468
[I 14:56:31.949 LabApp]  or http://127.0.0.1:8890/?token=d739f9a1a3539248da41cd364ba8b5eae380fca905393468
[I 14:56:31.949 LabApp] Use Control-C to stop this server and shut down all kernels (twice to skip confirmation).
[C 14:56:31.957 LabApp]

    To access the notebook, open this file in a browser:
        file:///Users/dradecic/Library/Jupyter/runtime/nbserver-85721-open.html
    Or copy and paste one of these URLs:
        http://localhost:8890/?token=d739f9a1a3539248da41cd364ba8b5eae380fca905393468
     or http://127.0.0.1:8890/?token=d739f9a1a3539248da41cd364ba8b5eae380fca905393468
[I 14:56:37.680 LabApp] Build is up to date
```

Figure 2.9 – Starting JupyterLab for standalone installation

10. For the final check, we will take a look at which Python version came with the environment. This can be done straight from the notebooks, as shown in the following figure:

```
[1]:  import sys

[2]:  sys.version

[2]:  '3.8.3 (default, Jul  2 2020, 11:26:31) \n[Clang 10.0.0 ]'
```

Figure 2.10 – Checking the Python version for the standalone installation

11. Finally, we will see whether the TPOT library was installed by importing it and printing the version. This check can also be done from the notebooks. Follow the instructions in the following figure to see how:

```
[7]:  import tpot

[8]:  tpot.__version__

[8]:  '0.11.6.post1'
```

Figure 2.11 – Checking the TPOT version for the standalone installation

TPOT is now successfully installed in a virtual environment. The next section covers how to install and configure the environment with Anaconda.

Installing and configuring TPOT through Anaconda

Before proceeding, make sure you have Anaconda installed on your machine. We will use Anaconda to create and manage our environment and do the configurations from there:

1. To start, open up Anaconda Navigator:

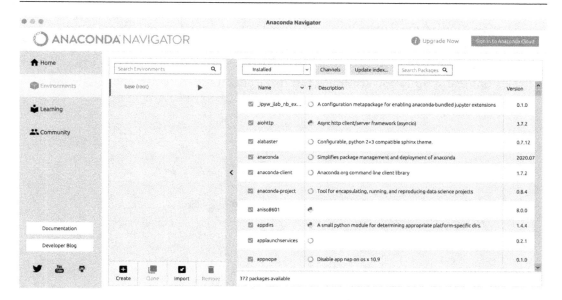

Figure 2.12 – Anaconda Navigator

2. To create a new virtual environment, click on the **Create** button in the bottom-left portion of the screen:

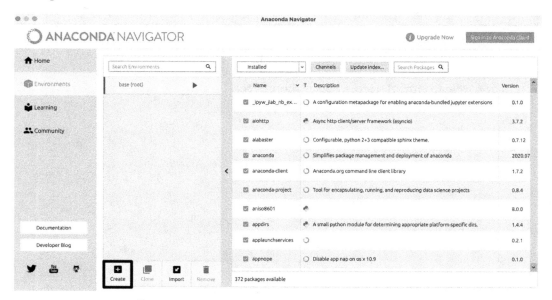

Figure 2.13 – Creating a new environment in Anaconda

3. After clicking on the **Create** button, a modal window will pop up. In this window, you will have to specify the environment name and Python version. Stick to the newest version available, which is 3.8 at this point. How you will name the environment is up to you arbitrary; just make sure it is descriptive enough for you to remember:

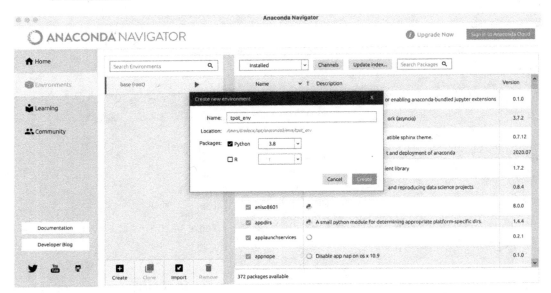

Figure 2.14 – Configuring the virtual environment name and Python version

After a couple of seconds, you will see the new environment listed below the `base (root)` environment. Here's how it should look:

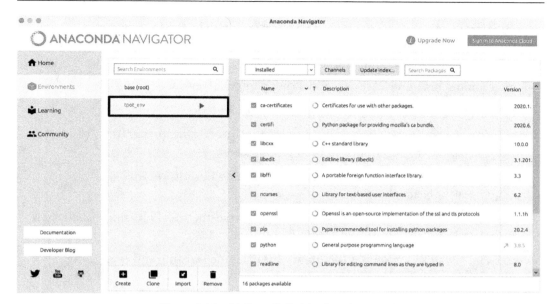

Figure 2.15 – Listing of all virtual environments

4. You are now ready to install libraries in your virtual environment. Anaconda makes it easy to open the environment from the terminal, by clicking on the play button and selecting the **Open Terminal** option. This is visible in the following figure:

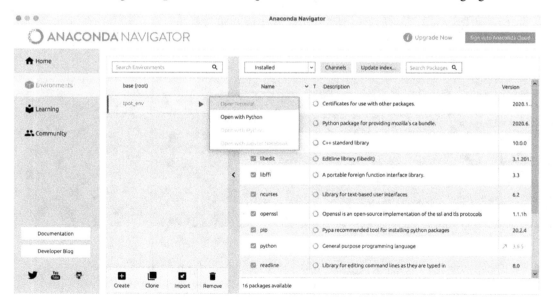

Figure 2.16 – Opening the virtual environment from the terminal

Once the terminal window opens up, you are ready to install the libraries. Throughout the entire book, we will need the following:

- `jupyterlab`: A notebook environment required for analyzing and exploring data and building machine learning models in an interactive way.

- `numpy`: Python's go-to library for numerical computations.

- `pandas`: A well-known library for data loading, processing, preparation, transformation, aggregation, and even visualization.

- `matplotlib`: Python's standard data visualization library. We will use it sometimes for basic plots.

- `seaborn`: A data visualization library with more aesthetically pleasing visuals than `matplotlib`.

- `scikit-learn`: Python's go-to library for machine learning and everything related to it.

- `TPOT`: Used to find optimal machine learning pipelines in an automated fashion.

5. To install every mentioned library, you can execute the following line from the opened terminal window:

```
> pip install jupyterlab numpy pandas matplotlib seaborn
  scikit-learn TPOT
```

Python should immediately start downloading and installing the libraries, as shown in the following figure:

```
(tpot_env) dradecic@Darios-MBP ~ % pip install jupyterlab numpy pandas matplotlib seaborn scikit-learn TPOT
Collecting jupyterlab
  Using cached jupyterlab-2.2.9-py3-none-any.whl (7.9 MB)
Collecting numpy
  Using cached numpy-1.19.4-cp38-cp38-macosx_10_9_x86_64.whl (15.3 MB)
Collecting pandas
  Using cached pandas-1.1.4-cp38-cp38-macosx_10_9_x86_64.whl (10.1 MB)
Collecting matplotlib
  Using cached matplotlib-3.3.3-cp38-cp38-macosx_10_9_x86_64.whl (8.5 MB)
Collecting seaborn
  Using cached seaborn-0.11.0-py3-none-any.whl (283 kB)
Collecting scikit-learn
  Using cached scikit_learn-0.23.2-cp38-cp38-macosx_10_9_x86_64.whl (7.2 MB)
Collecting TPOT
```

Figure 2.17 – Library installation through the terminal

6. To test whether the environment was successfully configured, we can open `JupyterLab` from the terminal. Execute the following shell command once the libraries are installed:

```
> jupyter lab
```

If you see something similar to the following, then everything went according to plan. The browser window with Jupyter should open immediately:

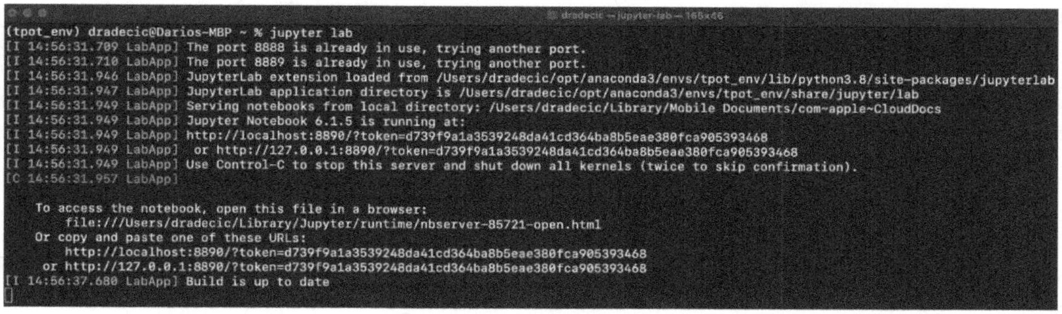

Figure 2.18 – Starting JupyterLab from the terminal

7. For the final check, we will take a look at which Python version came with the environment. This can be done straight from the notebooks, as shown in the following figure:

```
[1]:  import sys

[2]:  sys.version

[2]:  '3.8.5 (default, Sep  4 2020, 02:22:02) \n[Clang 10.0.0 ]'
```

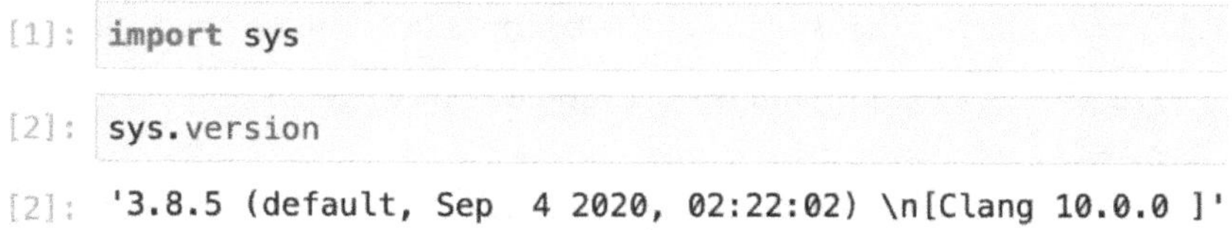

Figure 2.19 – Checking the Python version

8. Finally, we will see whether the TPOT library was installed by importing it and printing the version. This check can also be done from the notebooks. Follow the instructions in the following figure to see how:

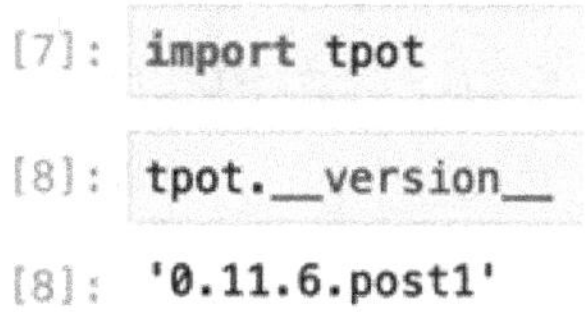

```
[7]:  import tpot

[8]:  tpot.__version__

[8]:  '0.11.6.post1'
```

Figure 2.20 – Checking the TPOT version

We are now ready to proceed with the practical uses of TPOT.

Summary

You've learned a lot in this chapter – from how TPOT works and GP to setting up the environment with `pip` and Anaconda. You are now ready to tackle hands-on tasks in an automated way.

The following chapter dives deep into handling regression tasks with TPOT with a couple of examples. Everything discussed during this chapter will become much clearer soon, after we get our hands dirty. Then, in *Chapter 4, Exploring before Classification*, you will further reinforce your knowledge by solving classification tasks.

Q&A

1. In your own words, define the TPOT library.

2. Name and explain a couple of TPOT 's limitations.

3. How would you limit the optimization time in TPOT?

4. Briefly define the term "genetic programming."

5. List and explain the five parameters of the `tpot.TPOTRegressor` class.

6. List and explain the different and new parameters introduced in the `tpot.TPOTClassifier` class.

7. What are virtual environments and why are they useful?

Further reading

Here are the sources we referenced in this chapter:

- *Genetic programming page*: `http://geneticprogramming.com`
- *TPOT documentation page*: `http://epistasislab.github.io/tpot/`

3
Exploring Regression with TPOT

In this chapter, you'll get hands-on experience with automated regression modeling through three datasets. You will learn how to handle regression tasks with TPOT in an automated manner with tons of practical examples, tips, and advice.

We will go through essential topics such as dataset loading, exploratory data analysis, and basic data preparation first. Then, we'll get our hands dirty with TPOT. You will learn how to train models in an automated way and how to evaluate those models.

Before training models automatically, we will see how good performance can be obtained with basic models, such as linear regression. These models will serve as a baseline that TPOT needs to outperform.

This chapter will cover the following topics:

- Applying automated regression modeling to the fish market dataset
- Applying automated regression modeling to the insurance dataset
- Applying automated regression modeling to the vehicle dataset

Technical requirements

To complete this chapter, you will need a computer with Python and TPOT installed. The previous chapter demonstrated how to set up the environment from scratch for both standalone Python installation and installation through Anaconda. Refer to *Chapter 2, Deep Dive into TPOT*, for detailed instructions on environment setup.

You can download the source code and datasets for this chapter here: `https://github.com/PacktPublishing/Machine-Learning-Automation-with-TPOT/tree/main/Chapter03`.

Applying automated regression modeling to the fish market dataset

This section demonstrates how to apply machine learning automation with TPOT to a regression dataset. The section uses the fish market dataset (`https://www.kaggle.com/aungpyaeap/fish-market`) for exploration and regression modeling. The goal is to predict the weight of a fish. You will learn how to load the dataset, visualize it, adequately prepare it, and how to find the best machine learning pipeline with TPOT:

1. The first thing to do is to load in the required libraries and load in the dataset. With regards to the libraries, you'll need `numpy`, `pandas`, `matplotlib`, and `seaborn`. Additionally, the `rcParams` module is imported with `matplotlib` to tweak the plot stylings a bit. You can find the code for this step in the following block:

    ```
    import numpy as np
    import pandas as pd
    import matplotlib.pyplot as plt
    import seaborn as sns
    from matplotlib import rcParams
    rcParams['axes.spines.top'] = False
    rcParams['axes.spines.right'] = False

    df = pd.read_csv('data/Fish.csv')
    df.head()
    ```

Here's how the first couple of rows look (the result from calling the `head()` method):

	Species	Weight	Length1	Length2	Length3	Height	Width
0	Bream	242.0	23.2	25.4	30.0	11.5200	4.0200
1	Bream	290.0	24.0	26.3	31.2	12.4800	4.3056
2	Bream	340.0	23.9	26.5	31.1	12.3778	4.6961
3	Bream	363.0	26.3	29.0	33.5	12.7300	4.4555
4	Bream	430.0	26.5	29.0	34.0	12.4440	5.1340

Figure 3.1 – First five rows of the fish market dataset

2. Exploratory data analysis comes in next. It's not a hard requirement for using TPOT, but you should always be aware of how your data looks. The first thing of interest is missing values. Here's how to check for them:

```
df.isnull().sum()
```

And here's the corresponding output:

```
Species      0
Weight       0
Length1      0
Length2      0
Length3      0
Height       0
Width        0
dtype: int64
```

Figure 3.2 – Count of missing values per column

As you can see, there are no missing values. This makes the data preparation process much easier and shorter.

3. The next step is to check how the target variable is distributed. For this dataset, we are trying to predict `Weight`. Here's the code for drawing a simple histogram:

```
plt.figure(figsize=(12, 7))
plt.title('Target variable (Weight) distribution',
size=20)
plt.xlabel('Weight', size=14)
plt.ylabel('Count', size=14)
plt.hist(df['Weight'], bins=15, color='#4f4f4f',
ec='#040404');
```

And here's how the histogram looks:

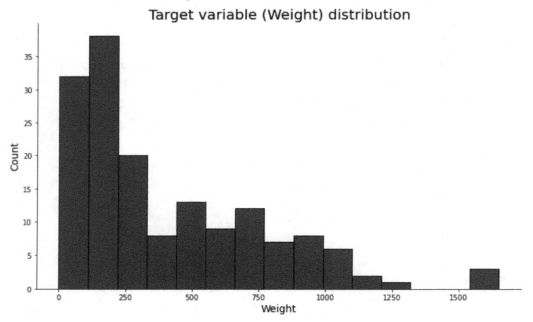

Figure 3.3 – Histogram of the target variable (Weight)

Most of the fish are light, but there are a couple of heavy ones present. Let's explore species further to get a better grasp.

4. The following code prints how many instances of a specific species there are (the number and percentage of the total), and also prints average and standard deviation for every attribute. To be more precise, a subset of the original dataset is kept where the species equals the specified species. Afterward, the number of records, total percentage, mean, and standard deviation are printed for every column in the subset.

This function is then called for every unique species:

```python
def describe_species(species):
    subset = df[df['Species'] == species]
    print(f'============ {species.upper()} ============')
    print(f'Count: {len(subset)}')
    print(f'Pct. total: {(len(subset) / len(df) *
100):.2f}%')
    for column in df.columns[1:]:
        avg = np.round(subset[column].mean(), 2)
        sd = np.round(subset[column].std(), 2)
        print(f'Avg. {column:>7}: {avg:6} +/- {sd:6}')

for species in df['Species'].unique():
    describe_species(species)
    print()
```

Here's the corresponding output:

```
============ BREAM ============      ============ PARKKI ============     ============ SMELT ============
Count: 35                            Count: 11                            Count: 14
Pct. total: 22.01%                   Pct. total: 6.92%                    Pct. total: 8.81%
Avg.  Weight: 617.83 +/- 209.21      Avg.  Weight: 154.82 +/-  78.76      Avg.  Weight:  11.18 +/-  4.13
Avg. Length1:  30.31 +/-    3.59     Avg. Length1:  18.73 +/-   3.28      Avg. Length1:  11.26 +/-  1.22
Avg. Length2:  33.11 +/-    3.91     Avg. Length2:  20.35 +/-   3.56      Avg. Length2:  11.92 +/-  1.43
Avg. Length3:  38.35 +/-    4.16     Avg. Length3:  22.79 +/-   3.96      Avg. Length3:  13.04 +/-  1.43
Avg.  Height:  15.18 +/-    1.96     Avg.  Height:   8.96 +/-   1.62      Avg.  Height:   2.21 +/-  0.35
Avg.   Width:   5.43 +/-    0.72     Avg.   Width:   3.22 +/-   0.64      Avg.   Width:   1.34 +/-  0.29

============ ROACH ============      ============ PERCH ============
Count: 20                            Count: 56
Pct. total: 12.58%                   Pct. total: 35.22%
Avg.  Weight: 152.05 +/-  88.83      Avg.  Weight: 382.24 +/- 347.62
Avg. Length1:  20.65 +/-   3.46      Avg. Length1:  25.74 +/-   8.56
Avg. Length2:  22.28 +/-   3.65      Avg. Length2:  27.89 +/-   9.02
Avg. Length3:  24.97 +/-   4.03      Avg. Length3:  29.57 +/-   9.53
Avg.  Height:   6.69 +/-   1.26      Avg.  Height:   7.86 +/-   2.88
Avg.   Width:   3.66 +/-   0.69      Avg.   Width:   4.75 +/-   1.77

============ WHITEFISH ============  ============ PIKE ============
Count: 6                             Count: 17
Pct. total: 3.77%                    Pct. total: 10.69%
Avg.  Weight: 531.0 +/- 309.6        Avg.  Weight: 718.71 +/- 494.14
Avg. Length1:  28.8 +/-   5.58       Avg. Length1:  42.48 +/-   9.03
Avg. Length2: 31.32 +/-   5.72       Avg. Length2:  45.48 +/-   9.71
Avg. Length3: 34.32 +/-   6.02       Avg. Length3:  48.72 +/-  10.17
Avg.  Height: 10.03 +/-   1.83       Avg.  Height:   7.71 +/-   1.66
Avg.   Width:  5.47 +/-   1.19       Avg.   Width:   5.09 +/-   1.14
```

Figure 3.4 – Feature exploration for every fish species

5. Finally, let's check for correlation between attributes. Correlation can be calculated only for numerical attributes. The following snippet shows you how to visualize a correlation matrix with the seaborn library:

```
plt.figure(figsize=(12, 9))
plt.title('Correlation matrix', size=20)
sns.heatmap(df.corr(), annot=True, cmap='Blues');
```

Here's the correlation matrix:

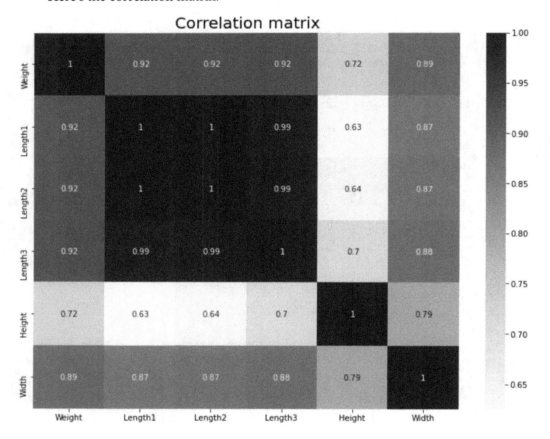

Figure 3.5 – Correlation matrix of features

You can do more in the exploratory data analysis process, but we'll stop here. This book shows you how to build automated models with TPOT, so we should spend most of the time there.

6.   There's one step left to do before modeling, and that is data preparation. We can't pass non-numerical attributes to the pipeline optimizer. We'll convert them to dummy variables for simplicity's sake and merge them with the original data afterward. Here's the code for doing so:

```
species_dummies = pd.get_dummies(df['Species'], drop_
first=True, prefix='Is')
df = pd.concat([species_dummies, df], axis=1)
df.drop('Species', axis=1, inplace=True)
df.head()
```

And here's how the dataset looks now:

	Is_Parkki	Is_Perch	Is_Pike	Is_Roach	Is_Smelt	Is_Whitefish	Weight	Length1	Length2	Length3	Height	Width
0	0	0	0	0	0	0	242.0	23.2	25.4	30.0	11.5200	4.0200
1	0	0	0	0	0	0	290.0	24.0	26.3	31.2	12.4800	4.3056
2	0	0	0	0	0	0	340.0	23.9	26.5	31.1	12.3778	4.6961
3	0	0	0	0	0	0	363.0	26.3	29.0	33.5	12.7300	4.4555
4	0	0	0	0	0	0	430.0	26.5	29.0	34.0	12.4440	5.1340

Figure 3.6 – First five rows of the fish market dataset after data preparation

As you can see, we deleted the Species column because it's not needed anymore. Let's begin with the modeling next.

7.   To start, we need to make a couple of imports and decide on the scoring strategy. TPOT comes with a couple of regression scoring metrics. The default one is neg_mean_squared_error. We can't escape the negative metric, but we can at least make it be in the same units as the target variable is. It makes no sense to predict weight and keep track of errors in weight squared. That's where **Root Mean Squared Error (RMSE)** comes into play. It is a simple metric that calculates the square root of the previously discussed mean squared error. Due to the square root operations, we're tracking errors in the original units (weight) instead of squared units (weight squared). We will define it with the help of lambda functions:

```
from tpot import TPOTRegressor
from sklearn.model_selection import train_test_split
from sklearn.metrics import mean_squared_error, make_
scorer

rmse = lambda y, y_hat: np.sqrt(mean_squared_error(y, y_
hat))
```

8. Next on the requirement list is the train test split. We will keep 75% of the data for training and evaluate on the rest:

```
X = df.drop('Weight', axis=1)
y = df['Weight']

X_train, X_test, y_train, y_test = train_test_split(
    X, y, test_size=0.25, random_state=42
)
```

Here's how many instances are in the train and test sets, respectively:

$$((119,), (40,))$$

Figure 3.7 – Number of instances in the training and test sets

9. Next, let's make a model with the linear regression algorithm. This model is just a baseline that TPOT needs to outperform:

```
from sklearn.linear_model import LinearRegression

lm = LinearRegression()
lm.fit(X_train, y_train)

lm_preds = lm.predict(X_test)
rmse(y_test, lm_preds)
```

Here's the corresponding RMSE value for linear regression on the test set:

$$82.10709519987495$$

Figure 3.8 – RMSE score for the linear regression model (baseline)

The baseline model is wrong by 82 units of weight on average. Not bad, considering we have weights up to 1,500.

10. Next, let's fit a TPOT pipeline optimization model. We will use our RMSE scorer and perform the optimization for 10 minutes. You can optimize for more time, but 10 minutes should outperform the baseline model:

```
rmse_scorer = make_scorer(rmse, greater_is_better=False)

pipeline_optimizer = TPOTRegressor(
    scoring=rmse_scorer,
    max_time_mins=10,
    random_state=42
)

pipeline_optimizer.fit(X_train, y_train)
```

After the optimization has finished, here's the output that's shown in the console:

```
TPOTRegressor(max_time_mins=10, random_state=42,
              scoring=make_scorer(<lambda>, greater_is_better=False))
```

Figure 3.9 – TPOT regressor output

11. Here's how to obtain the RMSE score:

```
pipeline_optimizer.score(X_test, y_test)
```

And here is the corresponding output:

−73.3463475920203

Figure 3.10 – RMSE score for TPOT optimized pipeline model

Don't worry about the minus sign before the number. The actual RMSE is 73.35 units of weight. The TPOT model outperformed the baseline one. That's all you need to know. TPOT gives us access to the best pipeline through the `fitted_pipeline_` attribute. Here's how it looks:

```
Pipeline(steps=[('minmaxscaler', MinMaxScaler()),
                ('stackingestimator',
                 StackingEstimator(estimator=SGDRegressor(alpha=0.0, eta0=1.0,
                                                          l1_ratio=0.0,
                                                          learning_rate='constant',
                                                          loss='huber',
                                                          penalty='elasticnet',
                                                          power_t=0.1,
                                                          random_state=42))),
                ('adaboostregressor',
                 AdaBoostRegressor(loss='exponential', n_estimators=100,
                                   random_state=42))])
```

Figure 3.11 – Full TPOT pipeline

12. As a final step, we can export the pipeline to a Python file. Here's how:

```
pipeline_optimizer.export('fish_pipeline.py')
```

Here's what the file looks like:

```
1  import numpy as np
2  import pandas as pd
3  from sklearn.ensemble import AdaBoostRegressor
4  from sklearn.linear_model import SGDRegressor
5  from sklearn.model_selection import train_test_split
6  from sklearn.pipeline import make_pipeline, make_union
7  from sklearn.preprocessing import MinMaxScaler
8  from tpot.builtins import StackingEstimator
9  from tpot.export_utils import set_param_recursive
10
11 # NOTE: Make sure that the outcome column is labeled 'target' in the data file
12 tpot_data = pd.read_csv('PATH/TO/DATA/FILE', sep='COLUMN_SEPARATOR', dtype=np.float64)
13 features = tpot_data.drop('target', axis=1)
14 training_features, testing_features, training_target, testing_target = \
15            train_test_split(features, tpot_data['target'], random_state=42)
16
17 # Average CV score on the training set was: -48.64388938943501
18 exported_pipeline = make_pipeline(
19     MinMaxScaler(),
20     StackingEstimator(estimator=SGDRegressor(alpha=0.0, eta0=1.0, fit_intercept=True,
21                                              l1_ratio=0.0, learning_rate="constant",
22                                              loss="huber", penalty="elasticnet",
23                                              power_t=0.1)),
24     AdaBoostRegressor(learning_rate=1.0, loss="exponential", n_estimators=100)
25 )
26 # Fix random state for all the steps in exported pipeline
27 set_param_recursive(exported_pipeline.steps, 'random_state', 42)
28
29 exported_pipeline.fit(training_features, training_target)
30 results = exported_pipeline.predict(testing_features)
```

Figure 3.12 – Source code of the TPOT pipeline

You can now use this file to make predictions on new, unseen data.

In this section, you've built your first automated machine learning pipeline with TPOT on a simple dataset. Most of the time, in practice, the steps you take will look similar. It's the data cleaning and preparation where things differ. Always make sure to prepare your dataset adequately before passing it to TPOT. Sure, TPOT does many things for you, but it can't turn garbage data into a usable model.

In the next section, you'll see how to apply TPOT to the medical insurance dataset.

Applying automated regression modeling to the insurance dataset

This section demonstrates how to apply an automated machine learning solution to a slightly more complicated dataset. You will use the medical insurance cost dataset (https://www.kaggle.com/mirichoi0218/insurance) to predict how much insurance will cost based on a couple of predictor variables. You will learn how to load the dataset, perform exploratory data analysis, how to prepare it, and how to find the best machine learning pipeline with TPOT:

1. As with the previous example, the first step is to load in the libraries and the dataset. We'll need `numpy`, `pandas`, `matplotlib`, and `seaborn` to start with the analysis. Here's how to import the libraries and load the dataset:

```
import numpy as np
import pandas as pd
import matplotlib.pyplot as plt
import seaborn as sns
from matplotlib import rcParams
rcParams['axes.spines.top'] = False
rcParams['axes.spines.right'] = False

df = pd.read_csv('data/insurance.csv')
df.head()
```

The first five rows are shown in the following figure:

	age	sex	bmi	children	smoker	region	charges
0	19	female	27.900	0	yes	southwest	16884.92400
1	18	male	33.770	1	no	southeast	1725.55230
2	28	male	33.000	3	no	southeast	4449.46200
3	33	male	22.705	0	no	northwest	21984.47061
4	32	male	28.880	0	no	northwest	3866.85520

Figure 3.13 – First five rows of the insurance dataset

2. We'll continue with the exploratory data analysis. As with the previous example, we'll first check for the number of missing values. Here's the code for doing so:

```
df.isnull().sum()
```

The following figure shows counts of missing values per column:

```
age         0
sex         0
bmi         0
children    0
smoker      0
region      0
charges     0
dtype: int64
```

Figure 3.14 – Missing value counts per column for the insurance dataset

As you can see, there are no missing values.

3. We're trying to predict the `charges` column with this dataset, so let's quickly check what type of values we can expect there. A histogram seems like an easy enough option. Here's the code needed for drawing one:

```
plt.figure(figsize=(12, 7))
plt.title('Target variable (charges) distribution',
size=20)
plt.xlabel('Charge', size=14)
plt.ylabel('Count', size=14)
plt.hist(df['charges'], bins=15, color='#4f4f4f',
ec='#040404');
```

And here's the resulting histogram:

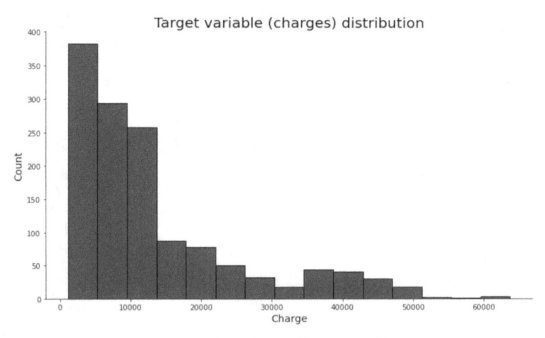

Figure 3.15 – Distribution of the target variable

So, values even go above $60,000.00. Most of them are lower, so it will be interesting to see how the model will handle it.

4. Let's dive deeper into the analysis and explore other variables. The goal is to see the average insurance costs for every categorical variable segment. We'll use the median as an average value, as it's less prone to outliers.

The easiest way to approach this analysis is to make a function that makes a bar chart for the specified column. The following function will come in handy for this example and many others in the future. It calculates a median from a grouped dataset and visualizes a bar chart with a title, labels, a legend, and text on top of the bars. You can use this function in general to visualize medians of some variable after a grouping operation is performed. It's best suited for categorical variables:

```python
def make_bar_chart(column, title, ylabel, xlabel, y_
offset=0.12, x_offset=700):
    ax = df.groupby(column).median()[['charges']].plot(
        kind='bar', figsize=(10, 6), fontsize=13,
color='#4f4f4f'
    )
    ax.set_title(title, size=20, pad=30)
    ax.set_ylabel(ylabel, fontsize=14)
    ax.set_xlabel(xlabel, fontsize=14)
    ax.get_legend().remove()

    for i in ax.patches:
        ax.text(i.get_x() + x_offset, i.get_height() + y_
offset, f'${str(round(i.get_height(), 2))}', fontsize=15)
    return ax
```

5. Let's now use this function to visualize the median insurance cost for smokers and non-smokers. Here's the code:

```python
make_bar_chart(
    column='smoker',
    title='Median insurance charges for smokers and
non-smokers',
    ylabel='Insurance charge ($)',
    xlabel='Do they smoke?',
    y_offset=700,
    x_offset=0.12
)
```

And here's the corresponding visualization:

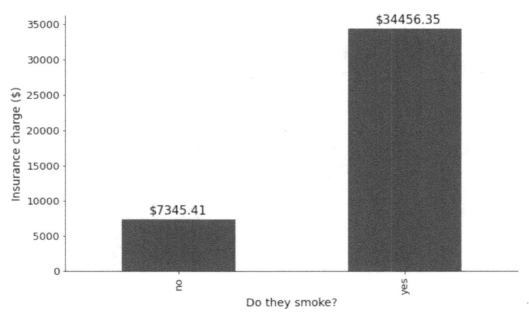

Figure 3.16 – Median insurance charges for smokers and non-smokers

As you can see, smokers pay an insurance fee several times higher than non-smokers.

6. Let's make a similar-looking visualization for comparing median insurance costs between genders:

```
make_bar_chart(
    column='sex',
    title='Median insurance charges between genders',
    ylabel='Insurance charge ($)',
    xlabel='Gender',
    y_offset=200,
    x_offset=0.15
)
```

You can see the visualization here:

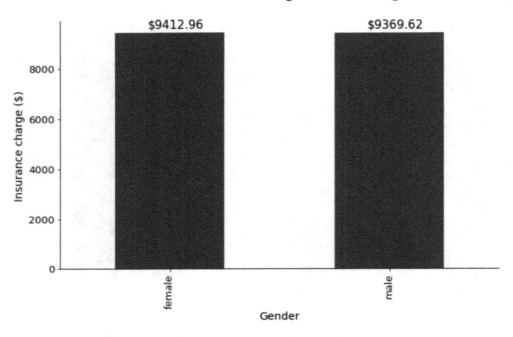

Figure 3.17 – Median insurance charges between genders

Not much of a difference here.

7. But what will happen if we compare median insurance costs by the number of children? The following code snippet does just that:

```
make_bar_chart(
    column='children',
    title='Median insurance charges by number of
children',
    ylabel='Insurance charge ($)',
    xlabel='Number of children',
    y_offset=200,
    x_offset=-0.15
)
```

Here's how the costs are distributed:

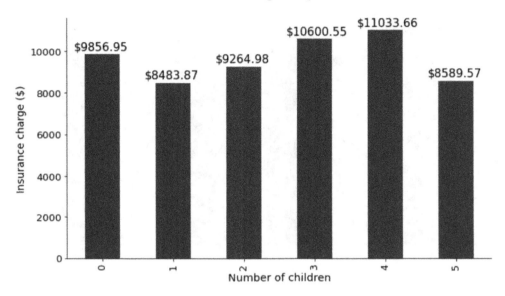

Figure 3.18 – Median insurance charges by number of children

The insurance costs seem to go up until the fifth child. Maybe there aren't that many families with five children. Can you confirm that on your own?

8. What about the region? Here's the code for visualizing median insurance costs by region:

```
make_bar_chart(
    column='region',
    title='Median insurance charges by region',
    ylabel='Insurance charge ($)',
    xlabel='Region',
    y_offset=200,
    x_offset=0
)
```

The cost distribution per region is shown in the following figure:

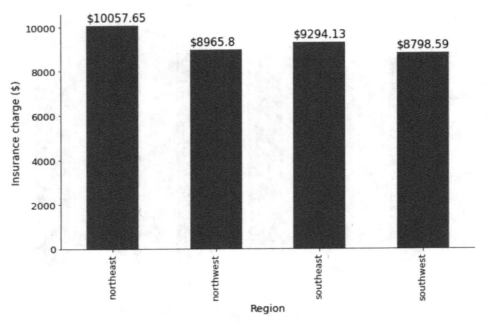

Figure 3.19 – Median insurance charges by region

The values don't differ that much.

We've made a decent amount of visualizations and explored the dataset. It's now time to prepare it and apply machine learning models.

9. There are a couple of things we need to do for this dataset to be machine learning ready. First, we'll have to remap string values to integers for the columns sex and smoker. Then, we'll need to create dummy variables for the region column. This step is necessary because TPOT can't understand raw textual data.

Here's the code snippet that does the necessary preparation:

```
df['sex'] = [1 if x == 'female' else 0 for x in
df['sex']]
df.rename(columns={'sex': 'is_female'}, inplace=True)

df['smoker'] = [1 if x == 'yes' else 0 for x in
df['smoker']]
region_dummies = pd.get_dummies(df['region'], drop_
first=True, prefix='region')

df = pd.concat([region_dummies, df], axis=1)
df.drop('region', axis=1, inplace=True)

df.head()
```

Calling the head() function results in the dataset shown in the following figure:

	region_northwest	region_southeast	region_southwest	age	is_female	bmi	children	smoker	charges
0	0	0	1	19	1	27.900	0	1	16884.92400
1	0	1	0	18	0	33.770	1	0	1725.55230
2	0	1	0	28	0	33.000	3	0	4449.46200
3	1	0	0	33	0	22.705	0	0	21984.47061
4	1	0	0	32	0	28.880	0	0	3866.85520

Figure 3.20 – Insurance dataset after preparation

10. The dataset is now ready for predictive modeling. Before we do so, let's check for variable correlations with the target variable. The following snippet draws the correlation matrix with annotations:

```
plt.figure(figsize=(12, 9))
plt.title('Correlation matrix', size=20)
sns.heatmap(df.corr(), annot=True, cmap='Blues');
```

The corresponding correlation matrix is shown in the following figure:

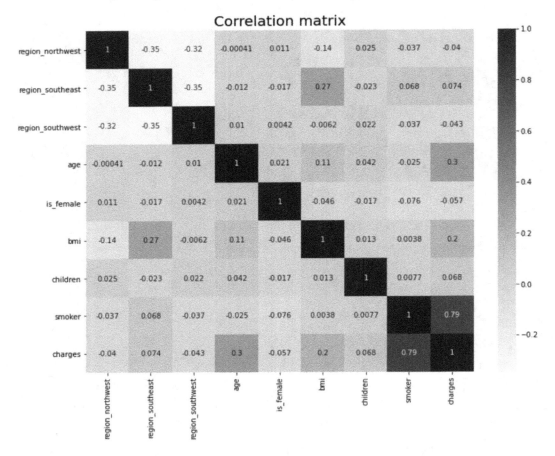

Figure 3.21 – Insurance dataset correlation matrix

Next stop – predictive modeling.

11. As before, the first step is to make a train/test split. The following code snippet shows you how to do that:

```
from sklearn.model_selection import train_test_split

X = df.drop('charges', axis=1)
y = df['charges']
```

```
X_train, X_test, y_train, y_test = train_test_split(
    X, y, test_size=0.25, random_state=42
)


y_train.shape, y_test.shape
```

The number of training and testing instances is shown in the following figure:

```
((1003,), (335,))
```

Figure 3.22 – Number of instances in train and test sets

12. We'll first make a baseline model with a linear regression algorithm. It will serve as something TPOT must outperform. You'll find a code snippet for training a baseline model here:

```
from sklearn.linear_model import LinearRegression
from sklearn.metrics import r2_score, mean_squared_error


rmse = lambda y, y_hat: np.sqrt(mean_squared_error(y, y_hat))


lm = LinearRegression()
lm.fit(X_train, y_train)


lm_preds = lm.predict(X_test)
print(f'R2   = {r2_score(y_test, lm_preds):.2f}')
print(f'RMSE = {rmse(y_test, lm_preds):.2f}')
```

Coefficient of determination (R2) and root mean squared error (RMSE) values are shown in the following figure:

```
R2   = 0.77
RMSE = 5926.02
```

Figure 3.23 – R2 and RMSE for the linear regression model

On average, a simple linear regression model is wrong by $5,926.02. This simple model captures 77% of the variance in the dataset.

13. We can further explore the linear regression model's feature importance by examining the assigned weights (coefficients).

The following code snippet prints the variable name and its corresponding coefficient:

```
for i, column in enumerate(df.columns[:-1]):
    coef = np.round(lm.coef_[i], 2)
    print(f'{column:17}: {coef:8}')
```

The output is shown in the following figure:

```
region_northwest :    -355.15
region_southeast :    -781.37
region_southwest :    -860.27
age              :     259.62
is_female        :     -45.62
bmi              :     339.81
children         :      426.5
smoker           :    23630.4
```

Figure 3.24 – Coefficients of a linear regression model

As you can see, the column with the largest coefficient is smoker. That makes sense, as it confirms our visualization made in the exploratory data analysis phase.

14. It's now time to bring in the big guns. We'll use the TPOT library to produce an automated machine learning pipeline. We'll optimize the pipeline for R2 score this time, but feel free to stick with RMSE or any other metric.

The following code snippet imports the TPOT library, instantiates it, and fits the pipeline:

```
from tpot import TPOTRegressor

pipeline_optimizer = TPOTRegressor(
    scoring='r2',
    max_time_mins=10,
    random_state=42,
    verbosity=2
)

pipeline_optimizer.fit(X_train, y_train)
```

After 10 minutes, you should see the following output in your notebook:

```
HBox(children=(FloatProgress(value=0.0, description='Optimization Progress', style=ProgressStyle(description_w…

Generation 1 - Current best internal CV score: 0.8506515769036094

Generation 2 - Current best internal CV score: 0.8506515769036094

Generation 3 - Current best internal CV score: 0.8506515769036094

Generation 4 - Current best internal CV score: 0.8506515769036094

Generation 5 - Current best internal CV score: 0.8508015012639911

Generation 6 - Current best internal CV score: 0.854533334580158

Generation 7 - Current best internal CV score: 0.854533334580158

10.01 minutes have elapsed. TPOT will close down.
TPOT closed during evaluation in one generation.
WARNING: TPOT may not provide a good pipeline if TPOT is stopped/interrupted in a early generation.

TPOT closed prematurely. Will use the current best pipeline.
```

Figure 3.25 – TPOT score per generation

The score on the training set started to increase in the last couple of generations. You'd likely get a slightly better model if you gave TPOT more time to train.

15. The R2 score on the test set can be obtained with the following code:

```
pipeline_optimizer.score(X_test, y_test)
```

The score is shown in the following figure:

0.8626764480217046

Figure 3.26 – TPOT R2 score on the test set

16. You can obtain R2 and RMSE values for the test set manually. The following code snippet shows you how:

```
tpot_preds = pipeline_optimizer.predict(X_test)

print(f'R2   = {r2_score(y_test, tpot_preds):.2f}')
print(f'RMSE = {rmse(y_test, tpot_preds):.2f}')
```

The corresponding scores are shown here:

```
R2   = 0.86
RMSE = 4552.02
```

Figure 3.27 – TPOT R2 and RMSE scores on the test set

17. As the last step, we'll export the optimized pipeline to a Python file. The following code snippet does it:

```
pipeline_optimizer.export('insurance_pipeline.py')
```

The Python code for the optimized pipeline is shown here:

```
1  import numpy as np
2  import pandas as pd
3  from sklearn.ensemble import ExtraTreesRegressor, GradientBoostingRegressor
4  from sklearn.model_selection import train_test_split
5  from sklearn.pipeline import make_pipeline, make_union
6  from sklearn.tree import DecisionTreeRegressor
7  from tpot.builtins import StackingEstimator
8  from tpot.export_utils import set_param_recursive
9
10 # NOTE: Make sure that the outcome column is labeled 'target' in the data file
11 tpot_data = pd.read_csv('PATH/TO/DATA/FILE', sep='COLUMN_SEPARATOR', dtype=np.float64)
12 features = tpot_data.drop('target', axis=1)
13 training_features, testing_features, training_target, testing_target = \
14             train_test_split(features, tpot_data['target'], random_state=42)
15
16 # Average CV score on the training set was: 0.854533334580158
17 exported_pipeline = make_pipeline(
18     StackingEstimator(estimator=DecisionTreeRegressor(
19         max_depth=2, min_samples_leaf=16, min_samples_split=9
20     )),
21     StackingEstimator(estimator=GradientBoostingRegressor(
22         alpha=0.75, learning_rate=1.0, loss="huber", max_depth=2,
23         max_features=0.7500000000000001, min_samples_leaf=5,
24         min_samples_split=16, n_estimators=100,
25         subsample=0.7500000000000001
26     )),
27     ExtraTreesRegressor(
28         bootstrap=False, max_features=0.8, min_samples_leaf=11,
29         min_samples_split=20, n_estimators=100
30     )
31 )
32 # Fix random state for all the steps in exported pipeline
33 set_param_recursive(exported_pipeline.steps, 'random_state', 42)
34
35 exported_pipeline.fit(training_features, training_target)
36 results = exported_pipeline.predict(testing_features)
```

Figure 3.28 – TPOT optimized pipeline for the insurance dataset

You can now use this file to make predictions on new, unseen data. It would be best to leave the pipeline to perform the optimization for as long as needed, but even 10 minutes was enough to produce good-quality models.

This section showed you how to build automated pipelines optimized for different metrics and with a bit more verbose output printed to the console. As you can see, the code for optimization is more or less identical. It's the data preparation that changes drastically from project to project, and that's where you'll spend most of your time.

In the next section, you'll see how to apply TPOT to the vehicle dataset.

Applying automated regression modeling to the vehicle dataset

This section shows how to develop an automated machine learning model on the most complex dataset thus far. You will use the vehicle dataset (https://www.kaggle.com/nehalbirla/vehicle-dataset-from-cardekho), so download it if you haven't already. The goal is to predict the selling price based on the various predictors, such as year made and kilometers driven.

This time, we won't focus on exploratory data analysis. You can do that on your own if you've followed the last two examples. Instead, we'll concentrate on dataset preparation and model training. There's a lot of work required to transform this dataset into something ready for machine learning, so let's get started right away:

1. Once again, the first step is to load in the libraries and the dataset. The requirements are the same as with previous examples. You'll need numpy, pandas, matplotlib, and seaborn. Here's how to import the libraries and load the dataset:

```
import numpy as np
import pandas as pd
import matplotlib.pyplot as plt
import seaborn as sns
from matplotlib import rcParams
rcParams['axes.spines.top'] = False
rcParams['axes.spines.right'] = False

df = pd.read_csv('data/Car.csv')
df.head()
```

Calling the `head()` function displays the first five rows. You can see how they look in the following figure:

	name	year	selling_price	km_driven	fuel	seller_type	transmission	
0	Maruti Swift Dzire VDI	2014	450000	145500	Diesel	Individual	Manual	First
1	Skoda Rapid 1.5 TDI Ambition	2014	370000	120000	Diesel	Individual	Manual	Second
2	Honda City 2017-2020 EXi	2006	158000	140000	Petrol	Individual	Manual	Third
3	Hyundai i20 Sportz Diesel	2010	225000	127000	Diesel	Individual	Manual	First
4	Maruti Swift VXI BSIII	2007	130000	120000	Petrol	Individual	Manual	First

Figure 3.29 – First five rows of the vehicle dataset

2. The dataset has a lot of columns, and not all of them are shown in *Figure 3.29*. The next step in the data preparation phase is to check for missing values. The following code snippet does that:

```
df.isnull().sum()
```

The results are shown in the following figure:

```
name                0
year                0
selling_price       0
km_driven           0
fuel                0
seller_type         0
transmission        0
owner               0
mileage           221
engine            221
max_power         215
torque            222
seats             221
dtype: int64
```

Figure 3.30 – Count of missing values for the vehicle dataset

Some of the values are missing, and we'll address this issue with the simplest approach – by removing them.

3. Removing missing values might not always be the best option. You should always investigate why the values are missing and if they can (or should) be somehow filled. This book focuses on machine learning automation, so we won't do that here.

Here's how you can drop the missing values:

```
df.dropna(inplace=True)
df.isnull().sum()
```

Executing the preceding code results in the following count:

```
name                0
year                0
selling_price       0
km_driven           0
fuel                0
seller_type         0
transmission        0
owner               0
mileage             0
engine              0
max_power           0
torque              0
seats               0
dtype: int64
```

Figure 3.31 – Removing missing values from the vehicle dataset

4. There are no missing values now, but that doesn't mean we're done with data preparation. Here's the list of steps required to make this dataset suitable for machine learning:

- Convert the `transmission` column to an integer – 1 if *manual*, 0 otherwise. Also, rename the column to `is_manual`.

- Remap the `owner` column to integers. Check the `remap_owner()` function for further clarifications.

- Extract car brand, mileage, engine, and max power from the corresponding attributes. The value of interest for all of the mentioned attributes is everything before the first space.

- Create dummy variables from the attributes `name`, `fuel`, and `seller_type`.

- Concatenate the original dataset with dummy variables and drop unnecessary attributes.

Here is the code for the `remap_owner()` function:

```
def remap_owner(owner):
    if owner == 'First Owner': return 1
    elif owner == 'Second Owner': return 2
    elif owner == 'Third Owner': return 3
```

```
elif owner == 'Fourth & Above Owner': return 4
else: return 0
```

And here is the code for performing all of the mentioned transformations:

```
df['transmission'] = [1 if x == 'Manual' else 0 for x in
df['transmission']]
df.rename(columns={'transmission': 'is_manual'},
inplace=True)

df['owner'] = df['owner'].apply(remap_owner)

df['name'] = df['name'].apply(lambda x: x.split()[0])
df['mileage'] = df['mileage'].apply(lambda x: x.split()
[0]).astype(float)
df['engine'] = df['engine'].apply(lambda x: x.split()
[0]).astype(int)
df['max_power'] = df['max_power'].apply(lambda x:
x.split()[0]).astype(float)

brand_dummies = pd.get_dummies(df['name'], drop_
first=True, prefix='brand')
fuel_dummies = pd.get_dummies(df['fuel'], drop_
first=True, prefix='fuel')
seller_dummies = pd.get_dummies(df['seller_type'], drop_
first=True, prefix='seller')

df.drop(['name', 'fuel', 'seller_type', 'torque'],
axis=1, inplace=True)
df = pd.concat([df, brand_dummies, fuel_dummies, seller_
dummies], axis=1)
```

After applying the transformations, the dataset looks like this:

	year	selling_price	km_driven	is_manual	owner	mileage	engine	max_power	seats	brand_Ashok	...
0	2014	450000	145500	1	1	23.40	1248	74.00	5.0	0	...
1	2014	370000	120000	1	2	21.14	1498	103.52	5.0	0	...
2	2006	158000	140000	1	3	17.70	1497	78.00	5.0	0	...
3	2010	225000	127000	1	1	23.00	1396	90.00	5.0	0	...
4	2007	130000	120000	1	1	16.10	1298	88.20	5.0	0	...

5 rows × 44 columns

Figure 3.32 – Prepared vehicle dataset

Data in this format can be passed to a machine learning algorithm. Let's do that next.

5. As always, we'll start with the train test split. The following code snippet shows you how to perform it on this dataset:

```
from sklearn.model_selection import train_test_split

X = df.drop('selling_price', axis=1)
y = df['selling_price']

X_train, X_test, y_train, y_test = train_test_split(
    X, y, test_size=0.25, random_state=42
)

y_train.shape, y_test.shape
```

You can see how many instances are in both sets in *Figure 3.33*:

$$((5929,), (1977,))$$

Figure 3.33 – Number of instances in train and test sets

As you can see, this is a much larger dataset than we had before.

6. We won't use your standard metrics for evaluating regression models (R2 and RMSE) this time. We'll use **Mean Absolute Percentage Error** (**MAPE**) instead. The MAPE metric measures the error of a predictive model as a percentage. The metric can be calculated as an average of the absolute difference between the actual and predicted values divided by the actual values. You can also optionally multiply this value by 100 to get the actual percentage. It's a great evaluation metric if there are no outliers (extremes) in the data. This metric isn't built into the `scikit-learn` library, so we'll have to implement it manually. Here's how:

```
def mape(y, y_hat):
    y, y_hat = np.array(y), np.array(y_hat)
    return np.mean(np.abs((y - y_hat) / y)) * 100
```

7. And now it's time to make a baseline model. Once again, it will be a linear regression model, evaluated on the test set with R2 and MAPE metrics. Here's the code for implementing the baseline model:

```
from sklearn.linear_model import LinearRegression
from sklearn.metrics import r2_score

lm = LinearRegression()
lm.fit(X_train, y_train)

lm_preds = lm.predict(X_test)
print(f'R2   = {r2_score(y_test, lm_preds):.2f}')
print(f'MAPE = {mape(y_test, lm_preds):.2f}')
```

The corresponding results are shown in the following figure:

```
R2   = 0.87
MAPE = 42.87
```

Figure 3.34 – R2 and MAPE for the baseline model

On average, the baseline model is wrong by 43%. It's a lot, but we have to start somewhere.

8. Let's take a look at the linear regression model coefficient to determine which features are important. Here's the code for obtaining coefficients:

```
for i, column in enumerate(df.columns[:-1]):
    coef = np.round(lm.coef_[i], 2)
    print(f'{column:20}: {coef:12}')
```

And here are the coefficients:

```
year                 :      43014.37   brand_Lexus          :     206214.99
selling_price        :         -0.99   brand_MG             :    -327026.98
km_driven            :      -93417.99  brand_Mahindra       :    -256695.95
is_manual            :      -29134.49  brand_Maruti         :     689295.55
owner                :       -3223.42  brand_Mercedes-Benz  :     -57724.47
mileage              :          51.82  brand_Mitsubishi     :    -371642.93
engine               :        6442.18  brand_Nissan         :          0.0
max_power            :       -6664.42  brand_Opel           :    -369176.64
seats                :          -0.0   brand_Renault        :    -422164.25
brand_Ashok          :      732079.41  brand_Skoda          :    -455033.96
brand_Audi           :     2062806.87  brand_Tata           :     -19784.95
brand_BMW            :     -413173.11  brand_Toyota         :    -432636.85
brand_Chevrolet      :      143836.25  brand_Volkswagen     :    1494271.37
brand_Daewoo         :     -456934.44  brand_Volvo          :     153467.18
brand_Datsun         :     -412771.18  fuel_Diesel          :     178327.49
brand_Fiat           :     -350143.14  fuel_LPG             :      58605.79
brand_Force          :      -351419.7  fuel_Petrol          :     -49419.66
brand_Ford           :     -359997.28  seller_Individual    :      -76786.2
brand_Honda          :      -350773.5
brand_Hyundai        :      382947.89
brand_Isuzu          :     1084736.28
brand_Jaguar         :      403502.72
brand_Jeep           :        61492.6
brand_Kia            :     2185633.63
brand_Land           :     3094325.97
```

Figure 3.35 – Baseline model coefficients

Just take a moment to appreciate how interpretable this is. The higher the year, the newer the car is, which results in a higher price. The more kilometers the vehicle has driven, the more the price decreases. It also looks like cars with automatic transmissions cost more. You get the point. Interpretability is something that linear regression offers. But it lacks accuracy. That's what TPOT will improve.

9. Let's fit a TPOT model next and optimize it for MAPE score. We'll train the model for 10 minutes on every available CPU core (indicated by `n_jobs=-1`):

```
from tpot import TPOTRegressor
from sklearn.metrics import make_scorer

mape_scorer = make_scorer(mape, greater_is_better=False)

pipeline_optimizer = TPOTRegressor(
    scoring=mape_scorer,
    max_time_mins=10,
    random_state=42,
    verbosity=2,
```

```
        n_jobs=-1
)


    pipeline_optimizer.fit(X_train, y_train)
```

The output you'll see after 10 minutes is shown in the following figure:

```
HBox(children=(FloatProgress(value=0.0, description='Optimization Progress', style=ProgressStyle(description_w…

Generation 1 - Current best internal CV score: -15.112509587258568

Generation 2 - Current best internal CV score: -15.112509587258568

11.19 minutes have elapsed. TPOT will close down.
TPOT closed during evaluation in one generation.
WARNING: TPOT may not provide a good pipeline if TPOT is stopped/interrupted in a early generation.
```

Figure 3.36 – Output of a TPOT optimization process

It looks like 10 minutes wasn't nearly enough for TPOT to give its best.

The resulting pipeline is shown in the following figure:

```
Pipeline(steps=[('featureunion',
                FeatureUnion(transformer_list=[('functiontransformer-1',
                                                FunctionTransformer(func=<function copy at 0x7fd819775940>)),
                                               ('functiontransformer-2',
                                                FunctionTransformer(func=<function copy at 0x7fd819775940>))])),
                ('gradientboostingregressor',
                GradientBoostingRegressor(alpha=0.8, loss='lad', max_depth=10,
                                          max_features=0.6000000000000001,
                                          min_samples_leaf=4,
                                          min_samples_split=5, random_state=42,
                                          subsample=0.8))])
```

Figure 3.37 – Best fitted pipeline after 10 minutes

10. And now the moment of truth – did the MAPE decrease? Here's the code to find out:

```
    tpot_preds = pipeline_optimizer.predict(X_test)


    print(f'R2   = {r2_score(y_test, tpot_preds):.2f}')
    print(f'MAPE = {mape(y_test, tpot_preds):.2f}')
```

The output is shown in the following figure:

```
R2   = 0.97
MAPE = 14.68
```

Figure 3.38 – R2 and MAPE for the TPOT optimized model

As you can see, TPOT decreased the error significantly and increased the goodness of fit (R2) simultaneously. Just as expected.

The final code-along section showed you how easy it is to train automated models on a more complex dataset. The procedure is more or less identical, depending on the metric you're optimizing for, but it's the data preparation phase that makes all the difference.

If you spend more time preparing and analyzing the data, and maybe removing some noisy data, you will get better results, guaranteed! That's mainly the case when a lot of columns contain text data. A lot of features can be extracted from there.

Summary

This was the first purely hands-on chapter in the book. You've connected the theory from the previous chapters with practice. You've built not one, but three fully automated machine learning models. Without any kind of doubt, you should now be able to use TPOT to solve any type of regression problem.

As with most things in data science and machine learning, 90% of the work boils down to data preparation. TPOT can make this percentage even higher because less time is spent designing and tweaking the models. Use this extra time wisely, and get yourself fully acquainted with the dataset. There's no way around it.

In the next chapter, you'll see how to build automated machine learning models for classification datasets. That chapter will also be entirely hands-on. Later, in *Chapter 5, Parallel Training with TPOT and Dask*, we'll combine both theory and practice.

Q&A

1. Which type of data visualization lets you explore the distribution of a continuous variable?

2. Explain R2, RMSE, and MAPE metrics.

3. Can you use a custom scoring function with TPOT? If yes, how?

4. Why is it essential to build baseline models first? Which algorithm is considered as a "baseline" for regression tasks?

5. What do the coefficients of a linear regression model tell you?

6. How do you use all CPU cores when training TPOT models?

7. Can you use TPOT to obtain the Python code of the best pipeline?

4
Exploring Classification with TPOT

In this chapter, you'll continue going through hands-on examples of automated machine learning. You will learn how to handle classification tasks with TPOT in an automated manner by going through three complete datasets.

We will cover essential topics such as dataset loading, cleaning, necessary data preparation, and exploratory data analysis. Then, we'll dive deep into classification with TPOT. You will learn how to train and evaluate automated classification models.

Before training models automatically, you will see how good models can be obtained with basic classification algorithms, such as logistic regression. This model will serve as the baseline that TPOT needs to outperform.

This chapter will cover the following topics:

- Applying automated classification modeling to the Iris dataset
- Applying automated classification modeling to the Titanic dataset

Technical requirements

To complete this chapter, You will need to have Python and TPOT installed in your computer with Python and TPOT installed. Refer to *Chapter 2, Deep Dive into TPOT*, for detailed instructions on environment setup. If the concept of classification is entirely new to you, refer to *Chapter 1, Machine Learning and the Idea of Automation*.

You can download the source code and dataset for this chapter here: `https://github.com/PacktPublishing/Machine-Learning-Automation-with-TPOT/tree/main/Chapter04`.

Applying automated classification models to the iris dataset

Let's start simple, with one of the most basic datasets out there – the Iris dataset (`https://en.wikipedia.org/wiki/Iris_flower_data_set`). The challenge here won't be to build an automated model but to build a model that can outperform the baseline model. The Iris dataset is so simple that even the most basic classification algorithm can achieve high accuracy.

Because of that, you should focus on getting the classification basics down in this section. You'll have enough time to worry about performance later:

1. As with the regression section, the first thing you should do is import the required libraries and load the dataset. You'll need numpy, pandas, matplotlib, and seaborn for starters. The matplotlib.rcParams module is imported to tweak the default stylings.

    Here's the code snippet for library imports and dataset loading:

    ```
    import numpy as np
    import pandas as pd
    import matplotlib.pyplot as plt
    import seaborn as sns
    from matplotlib import rcParams
    rcParams['axes.spines.top'] = False
    rcParams['axes.spines.right'] = False

    df = pd.read_csv('data/iris.csv')
    df.head()
    ```

And here is the output returned by the `head()` function:

	sepal_length	sepal_width	petal_length	petal_width	species
0	5.1	3.5	1.4	0.2	setosa
1	4.9	3.0	1.4	0.2	setosa
2	4.7	3.2	1.3	0.2	setosa
3	4.6	3.1	1.5	0.2	setosa
4	5.0	3.6	1.4	0.2	setosa

Figure 4.1 – Head of the Iris dataset

Great – just what we need to get started.

2. The next step is to check if data quality is good enough to be passed to a machine learning algorithm. The first step here is to check for missing values. The following code snippet does just that:

```
df.isnull().sum()
```

The output is shown in the following figure:

```
sepal_length    0
sepal_width     0
petal_length    0
petal_width     0
species         0
dtype: int64
```

Figure 4.2 – Missing value counts per column for the Iris dataset

There seem to be no missing values, so we can proceed.

3. Let's now check for class distribution in the target variable. This refers to the number of instances belonging to each class – `setosa`, `virginica`, and `versicolor`, in this case. Machine learning models are known to perform poorly if a severe class imbalance is present.

The following code snippet visualizes the class distribution:

```
ax = df.groupby('species').count().plot(kind='bar',
figsize=(10, 6), fontsize=13, color='#4f4f4f')
ax.set_title('Iris Dataset target variable distribution',
size=20, pad=30)
ax.set_ylabel('Count', fontsize=14)
ax.set_xlabel('Species', fontsize=14)
ax.get_legend().remove()
```

The visualization is shown in the following figure:

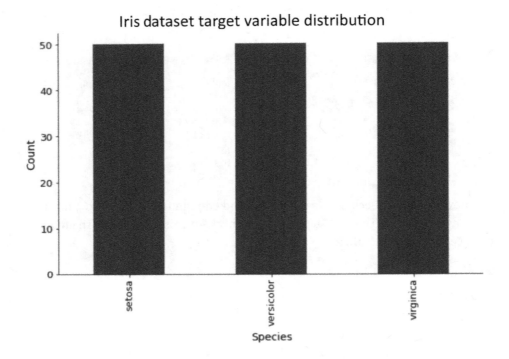

Figure 4.3 – Iris dataset target variable distribution

The Iris dataset is as nice as they come – so once again, nothing for us to do preparation-wise.

4. The final step in the data exploratory analysis and preparation is to check for correlation. A high correlation between features typically means there's some redundancy in the dataset, at least to a degree.

    The following code snippet plots a correlation matrix with annotations:

    ```
    plt.figure(figsize=(12, 9))
    plt.title('Correlation matrix', size=20)
    sns.heatmap(df.corr(), annot=True, cmap='Blues');
    ```

    The correlation matrix is shown in the following figure:

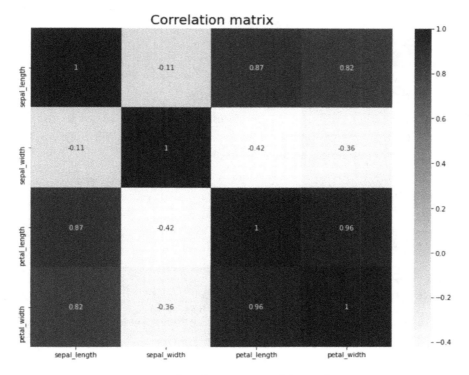

Figure 4.4 – Correlation matrix of the Iris dataset

As expected, there's a strong correlation between most of the features.

You're now familiar with the Iris dataset, which means we can move on to modeling the next.

5. Let's build a baseline model with a logistic regression algorithm first. It will serve as a starting model that TPOT needs to outperform.

The first step in the process is the train/test split. The following code snippet does just that, and it also prints the number of instances in both sets:

```
from sklearn.model_selection import train_test_split

X = df.drop('species', axis=1)
y = df['species']

X_train, X_test, y_train, y_test = train_test_split(
    X, y, test_size=0.25, random_state=3
)

y_train.shape, y_test.shape
```

The number of instances is shown in the following figure:

$$((112,), (38,))$$

Figure 4.5 – Number of instances in train and test sets

Let's build the baseline model next.

6.   As mentioned earlier, we'll use logistic regression for the job. The code snippet below fits a logistic regression model, makes the predictions on the test set, and prints a confusion matrix of actual and predicted values:

```
from sklearn.linear_model import LogisticRegression
from sklearn.metrics import confusion_matrix

lm = LogisticRegression(random_state=42)
lm.fit(X_train, y_train)

lm_preds = lm.predict(X_test)
print(confusion_matrix(y_test, lm_preds))
```

The corresponding confusion matrix is shown in the following figure:

```
[[15  0  0]
 [ 0 11  1]
 [ 0  0 11]]
```

Figure 4.6 – Logistic regression confusion matrix for the Iris dataset

As you can see, there's only one misclassification – a false positive for the *virginica* class. Even the baseline model performed exceptionally well. The code lines that follow obtain the accuracy for a baseline model:

```
from sklearn.metrics import accuracy_score

print(accuracy_score(y_test, lm_preds))
```

The accuracy score is shown in the following image:

0.9736842105263158

Figure 4.7 – Accuracy on the test set with logistic regression for the Iris dataset

And there you have it – 97% accuracy and only a single misclassification out of the box, with the simplest classification algorithm. Let's see if TPOT can outperform that next.

7. Let's build an automated classification model next. We'll optimize for accuracy and train for 10 minutes – similar to what we did in *Chapter 3, Exploring Regression with TPOT*. The code snippet below imports TPOT, instantiates a pipeline optimizer, and trains the model on the training datasets:

```
from tpot import TPOTClassifier

pipeline_optimizer = TPOTClassifier(
    scoring='accuracy',
    max_time_mins=10,
    random_state=42,
    verbosity=2
)


pipeline_optimizer.fit(X_train, y_train)
```

TPOT managed to fit 18 generations on my machine, which are shown in the following figure:

```
Generation 1 - Current best internal CV score: 0.9822134387351777

Generation 2 - Current best internal CV score: 0.9822134387351777

Generation 3 - Current best internal CV score: 0.9826086956521738

Generation 4 - Current best internal CV score: 0.9826086956521738

Generation 5 - Current best internal CV score: 0.9826086956521738

Generation 6 - Current best internal CV score: 0.9826086956521738

Generation 7 - Current best internal CV score: 0.9826086956521738

Generation 8 - Current best internal CV score: 0.9826086956521738

Generation 9 - Current best internal CV score: 0.9826086956521738

Generation 10 - Current best internal CV score: 0.9826086956521738

Generation 11 - Current best internal CV score: 0.9826086956521738

Generation 12 - Current best internal CV score: 0.9826086956521738

Generation 13 - Current best internal CV score: 0.9826086956521738

Generation 14 - Current best internal CV score: 0.9826086956521738

Generation 15 - Current best internal CV score: 0.9913043478260869

Generation 16 - Current best internal CV score: 0.9913043478260869

Generation 17 - Current best internal CV score: 0.9913043478260869

Generation 18 - Current best internal CV score: 0.9913043478260869
```

Figure 4.8 – Output of a TPOT pipeline optimization on the Iris dataset

8. Let's see if training an automated model managed to increase accuracy. You can use the following snippet to obtain the accuracy score:

```
tpot_preds = pipeline_optimizer.predict(X_test)

accuracy_score(y_test, tpot_preds)
```

The accuracy score is shown in the following figure:

0.9736842105263158

Figure 4.9 – Accuracy on the test set with an automated model for the Iris dataset

As you can see, the accuracy on the test set didn't improve. If you were to make a scatter plot of the target variable and features, you would see some overlap for the *virginica* and *versicolor* classes. That's most likely the case here, and no amount of training would manage to correctly classify this single instance.

9. There's only two things left to do here, and both are optional. The first one is to see what TPOT declared as an optimal pipeline after 10 minutes of training. The following code snippet will output that pipeline to the console:

```
pipeline_optimizer.fitted_pipeline_
```

The corresponding pipeline is shown in the following figure:

```
Pipeline(steps=[('nystroem',
                 Nystroem(gamma=0.30000000000000004, kernel='cosine',
                          n_components=3, random_state=42)),
                ('kneighborsclassifier',
                 KNeighborsClassifier(n_neighbors=6, p=1, weights='distance'))])
```

Figure 4.10 – Optimal TPOT pipeline for the Iris dataset

10. As always, you can also export the pipeline with the export() function:

```
pipeline_optimizer.export('iris_pipeline.py')
```

The entire Python code is shown in the following figure:

```
1   import numpy as np
2   import pandas as pd
3   from sklearn.kernel_approximation import Nystroem
4   from sklearn.model_selection import train_test_split
5   from sklearn.neighbors import KNeighborsClassifier
6   from sklearn.pipeline import make_pipeline
7   from tpot.export_utils import set_param_recursive
8
9   # NOTE: Make sure that the outcome column is labeled 'target' in the data file
10  tpot_data = pd.read_csv('PATH/TO/DATA/FILE', sep='COLUMN_SEPARATOR', dtype=np.float64)
11  features = tpot_data.drop('target', axis=1)
12  training_features, testing_features, training_target, testing_target = \
13              train_test_split(features, tpot_data['target'], random_state=42)
14
15  # Average CV score on the training set was: 0.9913043478260869
16  exported_pipeline = make_pipeline(
17      Nystroem(gamma=0.30000000000000004, kernel="cosine", n_components=3),
18      KNeighborsClassifier(n_neighbors=6, p=1, weights="distance")
19  )
20  # Fix random state for all the steps in exported pipeline
21  set_param_recursive(exported_pipeline.steps, 'random_state', 42)
22
23  exported_pipeline.fit(training_features, training_target)
24  results = exported_pipeline.predict(testing_features)
```

Figure 4.11 – Python code for an optimal TPOT pipeline for the Iris dataset

And there you have it – your first fully automated classification model with TPOT. Yes, the dataset was as basic as they come, but the principle always remains the same. We'll make automated models on a more complex dataset next, so there will be time to get your hands dirty.

Applying automated classification modeling to the titanic dataset

We're now going to apply automated TPOT classification modeling to a slightly more complicated dataset. You'll get your hands dirty with the Titanic dataset (https://gist.githubusercontent. com/michhar/2dfd2de0d4f8727f873422c5d959fff5/raw/ fa71405126017e6a37bea592440b4bee94bf7b9e/titanic.csv) – a dataset containing various attributes and descriptions of passengers who did and did not survive the Titanic accident.

The goal is to build an automated model capable of predicting whether a passenger would have survived the accident, based on various input features, such as passenger class, gender, age, cabin, number of siblings, spouses, parents, and children, among other features.

We'll start by loading the libraries and the dataset next:

1. As always, the first step is to load in the libraries and the dataset. You'll need numpy, pandas, matplotlib, and seaborn to get you started. The Matplotlib. rcParams module is also imported, just to make the visualizations a bit more appealing.

    The following code snippet imports the libraries, loads in the dataset, and displays the first five rows:

    ```
    import numpy as np
    import pandas as pd
    import matplotlib.pyplot as plt
    from matplotlib import rcParams
    rcParams['axes.spines.top'] = False
    rcParams['axes.spines.right'] = False


    df = pd.read_csv('data/titanic.csv')
    df.head()
    ```

    Calling the head() function returns the first five rows of the dataset. They are shown in the following figure:

	PassengerId	Survived	Pclass	Name	Sex	Age	SibSp	Parch	Ticket	Fare	Cabin	Embarked
0	1	0	3	Braund, Mr. Owen Harris	male	22.0	1	0	A/5 21171	7.2500	NaN	S
1	2	1	1	Cumings, Mrs. John Bradley (Florence Briggs Th...	female	38.0	1	0	PC 17599	71.2833	C85	C
2	3	1	3	Heikkinen, Miss. Laina	female	26.0	0	0	STON/O2. 3101282	7.9250	NaN	S
3	4	1	1	Futrelle, Mrs. Jacques Heath (Lily May Peel)	female	35.0	1	0	113803	53.1000	C123	S
4	5	0	3	Allen, Mr. William Henry	male	35.0	0	0	373450	8.0500	NaN	S

    Figure 4.12 – Head of the Titanic dataset

    You can now proceed with the exploratory data analysis and preparation.

2. The first step in the exploratory data analysis and preparation is to check for missing values. The following code line does just that:

    ```
    df.isnull().sum()
    ```

    The preceding line of code reports back the number of missing values per column in the dataset, as shown in the following figure:

```
PassengerId       0
Survived          0
Pclass            0
Name              0
Sex               0
Age             177
SibSp             0
Parch             0
Ticket            0
Fare              0
Cabin           687
Embarked          2
```

Figure 4.13 – Missing values count per column for the Titanic dataset

As you can see, there are a lot of missing values present in the dataset. Most of the missing values are in the `Age` and `Cabin` attributes. It's easy to understand for `Cabin` – the value is missing if the passenger didn't have their own cabin.

We'll deal with these missing values later, but for now, let's shift our focus to data visualization, so you can better understand the dataset.

3. To avoid code duplication, let's define a single function for displaying a bar chart. The function shows a bar chart with column counts on top of the bars. It also allows you to specify for which dataset column you want to draw a bar chart, values for the title, *x*-axis label, and *y*-axis label, and also offsets for the counts.

    You can find the code for this function here:

```
def make_bar_chart(column, title, ylabel, xlabel, y_
offset=10, x_offset=0.2):
    ax = df.groupby(column).count()[['PassengerId']].
plot(
        kind='bar', figsize=(10, 6), fontsize=13,
color='#4f4f4f'
    )
    ax.set_title(title, size=20, pad=30)
    ax.set_ylabel(ylabel, fontsize=14)
    ax.set_xlabel(xlabel, fontsize=14)
    ax.get_legend().remove()

    for i in ax.patches:
        ax.text(i.get_x() + x_offset, i.get_height() + y_
offset, i.get_height(), fontsize=20)
    return ax
```

You'll use this function extensively during the next couple of pages. The goal is to visualize how categorical variables are distributed, so you can get a better understanding of the dataset.

4. To start, let's visualize how many passengers have survived and how many haven't. The previously declared `make_bar_chart()` function comes in handy for the job.

The following code snippet makes the visualization:

```
make_bar_chart(
    column='Survived',
    title='Distribution of the Survived variable',
    ylabel='Count',
    xlabel='Has the passenger survived? (0 = No, 1 =
Yes)'
);
```

The visualization is displayed in the following figure:

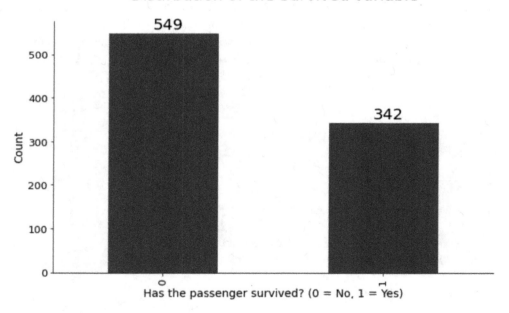

Figure 4.14 – Target class distribution for the Titanic dataset

As you can see, most of the passengers didn't survive the Titanic accident. This information alone doesn't tell you much because you don't know how many passengers survived per gender, passenger class, and other attributes.

You can use the `make_bar_chart()` function to make this type of visualization.

5. Let's continue our data visualization journey by visualizing the number of passengers in each passenger class. You can use the same `make_bar_chart()` function for this visualization. Just make sure to change the parameters accordingly.

The following code snippet visualizes the number of passengers per passenger class. The lower the class number, the better – a more expensive ticket, better service, and who knows, maybe a higher chance of survival:

```
make_bar_chart(
    column='Pclass',
    title='Distirbution of the Passenger Class variable',
    ylabel='Count',
    xlabel='Passenger Class (smaller is better)',
    x_offset=0.15
);
```

The visualization is shown in the following figure:

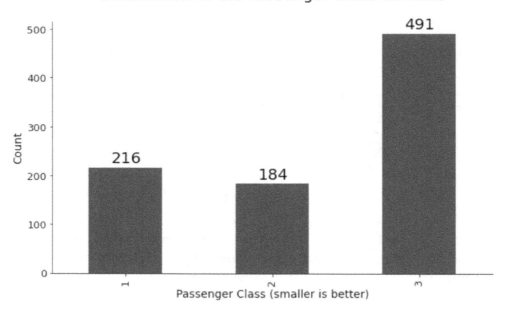

Figure 4.15 – Number of passengers per passenger class

As you can see, most of the passengers belong to the third class. This is expected, as there were more workers on board than rich people.

6. For the next step in the data visualization phase, let's see how the `Sex` attribute is distributed. This will give us insight into whether there were more women or men on board and how large the difference was.

    The following code snippet makes the visualization:

```
make_bar_chart(
    column='Sex',
    title='Distirbution of the Sex variable',
    ylabel='Count',
    xlabel='Gender'
);
```

    The visualization is shown in the following figure:

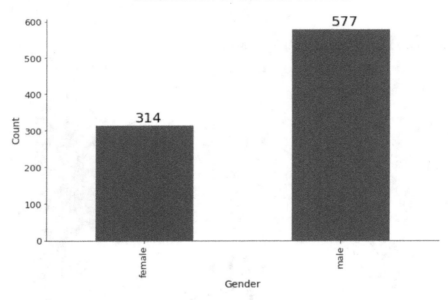

Figure 4.16 – Number of passengers per gender

    As you can see, there were definitely more men aboard. This is connected with the conclusion made in the previous visualization, where we concluded that there were many workers on board.

    Most of the workers are male, so this visualization makes sense.

7. Let's take a little break from the bar charts and visualize a continuous variable for change. The goal is to make a histogram of the `Fare` attribute, which will show the distribution of the amounts paid for the ticket.

The following code snippet draws a histogram for the mentioned attribute:

```
plt.figure(figsize=(12, 7))
plt.title('Fare cost distribution', size=20)
plt.xlabel('Cost', size=14)
plt.ylabel('Count', size=14)
plt.hist(df['Fare'], bins=15, color='#4f4f4f',
ec='#040404');
```

The histogram is shown in the following figure:

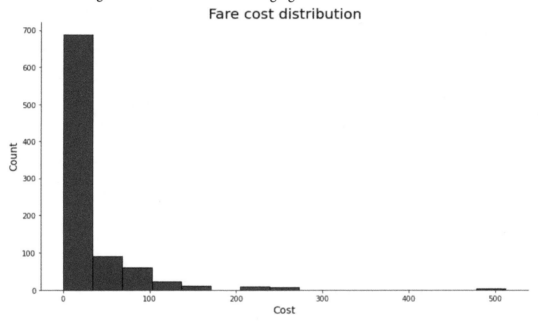

Figure 4.17 – Distribution of the Fare variable

It looks like most of the passengers paid 30 dollars or less for a ticket. As always, there are extreme cases. It seems like a single passenger paid around 500 dollars for the trip. Not a wise decision, taking into consideration how things ended.

8. Let's do something a bit different now. The Name attribute is more or less useless in this format. But if you take a closer look, you can see that every value in the mentioned attribute is formatted identically.

This means we can keep the single word after the first comma and store it in a new variable. We'll call this variable Title because it represents passenger titles (for example, Mr., Miss., and so on).

The following code snippet extracts the Title value to a new attribute and uses the `make_bar_chart()` function to visually represent different titles among Titanic passengers:

```
df['Title'] = df['Name'].apply(lambda x: x.split(',')[1].
strip().split(' ')[0])

make_bar_chart(
    column='Title',
    title='Distirbution of the Passenger Title variable',
    ylabel='Count',
    xlabel='Title',
    x_offset=-0.2
);
```

The results are shown in the following figure:

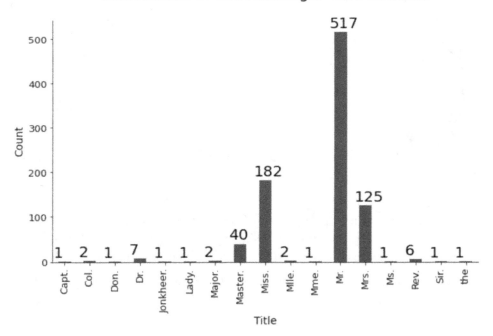

Figure 4.18 – Distribution of the passenger titles

Once again, these are expected results. Most of the passengers have common titles, such as Mr. and Miss. There's just a handful of them with unique titles. You could leave this column as is or turn it into a binary column – the value is zero if a title is common, and one otherwise. You'll see how to do that next.

9. That's about enough with regards to the exploratory data analysis. We've made quite a few visualizations, but you can always make more on your own.

It's now time to prepare the dataset for machine learning. The steps are described here:

a) Drop the columns that are of no use – `Ticket` and `PassengerId`. The first one is just a collection of dummy letters and numbers and is of no use for predictive modeling. The second one is an arbitrary ID, most likely generated with a database sequence. You can remove both by calling the `drop()` function.

b) Remap values in the `Sex` attribute to integers. The textual values *male* and *female* can't be passed to a machine learning algorithm directly. Some form of conversion is a must – so replace males with 0 and females with 1. The `replace()` function is the perfect candidate for the job.

c) Use the previously generated `Title` column and convert it into a binary one – the value is zero if the title is common (for example, Mr., Miss., and Mrs.) and one otherwise. You can then rename the column to something a bit more appropriate, such as `Title_Unusal`. The `Name` column isn't needed anymore, so delete it.

d) Handle missing values in the `Cabin` column by turning this attribute into a binary one – the value is zero if the value for the cabin is missing, and one otherwise. Name this new column `Cabin_Known`. After that, you can delete the `Cabin` column because it's not needed anymore, and it can't be passed to a machine learning model.

e) Create dummy variables with the `Embarked` attribute. This attribute indicates the port on which the passengers entered the ship. You be the judge of whether this attribute is even necessary, but we'll keep it for TPOT to decide. After declaring dummy variables, concatenate them to the original dataset and delete the `Embarked` column.

f) Handle missing values in the `Age` attribute somehow. There are many sophisticated methods, such as *KNN imputing* or *MissForest imputing*, but for simplicity's sake, just impute the missing values with a simple average.

The following code snippet shows you how to apply all of the mentioned transformations:

```
df.drop(['Ticket', 'PassengerId'], axis=1, inplace=True)

gender_mapper = {'male': 0, 'female': 1}
df['Sex'].replace(gender_mapper, inplace=True)
```

```
df['Title'] = [0 if x in ['Mr.', 'Miss.', 'Mrs.'] else 1
for x in df['Title']]
df = df.rename(columns={'Title': 'Title_Unusual'})
df.drop('Name', axis=1, inplace=True)

df['Cabin_Known'] = [0 if str(x) == 'nan' else 1 for x in
df['Cabin']]
df.drop('Cabin', axis=1, inplace=True)

emb_dummies = pd.get_dummies(df['Embarked'], drop_
first=True, prefix='Embarked')
df = pd.concat([df, emb_dummies], axis=1)
df.drop('Embarked', axis=1, inplace=True)

df['Age'] = df['Age'].fillna(int(df['Age'].mean()))

df.head()
```

You can take a peek at the prepared dataset by examining the following figure:

	Survived	Pclass	Sex	Age	SibSp	Parch	Fare	Title_Unusual	Cabin_Known	Embarked_Q	Embarked_S
0	0	3	0	22.0	1	0	7.2500	0	0	0	1
1	1	1	1	38.0	1	0	71.2833	0	1	0	0
2	1	3	1	26.0	0	0	7.9250	0	0	0	1
3	1	1	1	35.0	1	0	53.1000	0	1	0	1
4	0	3	0	35.0	0	0	8.0500	0	0	0	1

Figure 4.19 – Prepared Titanic dataset

And that's all you have to do with regard to data preparation. Scaling/standardization is not required, as TPOT will decide whether that step is necessary.

We'll begin with predictive modeling shortly – just one step remains.

10. Before you can train a classification model, you'll have to split the dataset into training and testing subsets. Keep in mind the `random_state` parameter – use the same value if you want the same data split:

```
from sklearn.model_selection import train_test_split

X = df.drop('Survived', axis=1)
y = df['Survived']

X_train, X_test, y_train, y_test = train_test_split(
```

```
    X, y, test_size=0.25, random_state=42
)


y_train.shape, y_test.shape
```

The last code line prints the number of instances in training and testing subsets. You can see the numbers in the following figure:

$$((668,), (223,))$$

Figure 4.20 – Number of instances in training and testing sets (Titanic)

Now you're ready to train predictive models.

11. Let's start with a baseline model – logistic regression. We'll train it on the train set and evaluate it on the test set. The following code snippet trains the model and prints the confusion matrix:

```
from sklearn.linear_model import LogisticRegression
from sklearn.metrics import confusion_matrix


lm = LogisticRegression(random_state=42)
lm.fit(X_train, y_train)


lm_preds = lm.predict(X_test)
print(confusion_matrix(y_test, lm_preds))
```

You can see the confusion matrix in the following figure:

```
[[111  23]
 [ 23  66]]
```

Figure 4.21 – Logistic regression confusion matrix (Titanic)

It looks like there's the same number of false positives and false negatives (23). If we take ratios into account, there are more false negatives. In translation, the baseline model is more likely to say that the passenger survived even if they didn't.

12. Interpreting the confusion matrix is great, but what if you want to look at a concrete number instead? Since this is a classification problem, you could use accuracy. But there's a "better" metric – **F1 score**. The value for this metric ranges between 0 and 1 (higher is better) and represents a harmonic mean between precision and recall.

Here's how to calculate it with Python:

```
from sklearn.metrics import f1_score


print(f1_score(y_test, lm_preds))
```

The value of the F1 score on the test set is shown in the following figure:

0.7415730337078652

Figure 4.22 – Logistic regression F1 score on the test set (Titanic)

The value of 0.74 isn't bad for a baseline model. Can TPOT outperform it? Let's train an automated model and see what happens.

13. In a similar fashion as before, we'll train an automated classification model for 10 minutes. Instead of accuracy, we'll optimize for the F1 score. By doing so, we can compare the F1 scores of an automated model with the baseline one.

The following code snippet trains the model on the training set:

```
from tpot import TPOTClassifier

pipeline_optimizer = TPOTClassifier(
    scoring='f1',
    max_time_mins=10,
    random_state=42,
    verbosity=2
)

pipeline_optimizer.fit(X_train, y_train)
```

In the following figure, you can see the output printed in the notebook during training. TPOT managed to train for 7 generations in 10 minutes, and the score increases as the model is training:

Generation 1 - Current best internal CV score: 0.7668883546121398

Generation 2 - Current best internal CV score: 0.7668883546121398

Generation 3 - Current best internal CV score: 0.7668883546121398

Generation 4 - Current best internal CV score: 0.7706865623331014

Generation 5 - Current best internal CV score: 0.7780316553709037

Generation 6 - Current best internal CV score: 0.7780316553709037

Generation 7 - Current best internal CV score: 0.7780316553709037

Figure 4.23 – TPOT pipeline optimization output (Titanic)

You are free to leave the model to train for longer than 10 minutes. Still, this time frame should be enough to outperform the baseline model.

14. Let's take a look at the value of the F1 score on the test set now. Remember, anything above 0.7415 means TPOT outperformed the baseline model.

    The following code snippet prints the F1 score:

    ```
    pipeline_optimizer.score(X_test, y_test)
    ```

    The corresponding F1 score is shown in the following figure:

    ```
    0.7701149425287356
    ```

    Figure 4.24 – TPOT optimized model F1 score on the test set (Titanic)

    It looks like TPOT outperformed the baseline model – as expected.

15. In case you're more trustworthy of basic metrics, such as accuracy, here's how you can compare it between baseline and automated models:

    ```
    tpot_preds = pipeline_optimizer.predict(X_test)

    from sklearn.metrics import accuracy_score

    print(f'Baseline model accuracy: {accuracy_score(y_test,
    lm_preds)}')
    print(f'TPOT model accuracy: {accuracy_score(y_test,
    tpot_preds)}')
    ```

    Corresponding accuracy scores are shown in the following figure:

    ```
    Baseline model accuracy: 0.7937219730941704
    TPOT model accuracy: 0.820627802690583
    ```

    Figure 3.25 – Accuracies of the baseline model and TPOT optimized model on the test set (Titanic)

    As you can see, the simple accuracy metric tells a similar story – the model built by TPOT is still better than the baseline one.

16. We are near the end of this practical example. There are two optional things left to do. The first one is to take a look at the optimal pipeline. You can obtain it with the following line of code:

    ```
    pipeline_optimizer.fitted_pipeline_
    ```

The optimal pipeline is shown in the following figure:

```
Pipeline(steps=[('featureunion',
                FeatureUnion(transformer_list=[('functiontransformer-1',
                                                FunctionTransformer(func=<function copy at 0x7fef9d7ba8b0>)),
                                               ('functiontransformer-2',
                                                FunctionTransformer(func=<function copy at 0x7fef9d7ba8b0>))])),
                ('xgbclassifier',
                 XGBClassifier(base_score=0.5, booster='gbtree',
                               colsample_bylevel=1, colsample_bynode=1,
                               colsample_bytree...id=-1,
                               importance_type='gain',
                               interaction_constraints='', learning_rate=0.1,
                               max_delta_step=0, max_depth=9,
                               min_child_weight=3, missing=nan,
                               monotone_constraints='()', n_estimators=100,
                               n_jobs=1, nthread=1, num_parallel_tree=1,
                               random_state=42, reg_alpha=0, reg_lambda=1,
                               scale_pos_weight=1, subsample=0.5,
                               tree_method='exact', validate_parameters=1,
                               verbosity=None))])
```

Figure 4.26 – TPOT optimized pipeline (Titanic)

As you can see, TPOT used extreme gradient boosting to solve this classification problem.

17. Finally, you can convert the optimal pipeline into Python code. Doing so makes the process of sharing the code that much easier. You can find the code for doing so here:

```
pipeline_optimizer.export('titanic_pipeline.py')
```

The full source code for the automated pipeline is shown in the following figure:

```
1  import numpy as np
2  import pandas as pd
3  from sklearn.model_selection import train_test_split
4  from sklearn.pipeline import make_pipeline, make_union
5  from tpot.builtins import StackingEstimator
6  from xgboost import XGBClassifier
7  from tpot.export_utils import set_param_recursive
8  from sklearn.preprocessing import FunctionTransformer
9  from copy import copy
10
11 # NOTE: Make sure that the outcome column is labeled 'target' in the data file
12 tpot_data = pd.read_csv('PATH/TO/DATA/FILE', sep='COLUMN_SEPARATOR', dtype=np.float64)
13 features = tpot_data.drop('target', axis=1)
14 training_features, testing_features, training_target, testing_target = \
15            train_test_split(features, tpot_data['target'], random_state=42)
16
17 # Average CV score on the training set was: 0.7780316553709037
18 exported_pipeline = make_pipeline(
19     make_union(
20         FunctionTransformer(copy),
21         FunctionTransformer(copy)
22     ),
23     XGBClassifier(learning_rate=0.1, max_depth=9, min_child_weight=3, n_estimators=100, nthread=1, subsample=0.5)
24 )
25 # Fix random state for all the steps in exported pipeline
26 set_param_recursive(exported_pipeline.steps, 'random_state', 42)
27
28 exported_pipeline.fit(training_features, training_target)
29 results = exported_pipeline.predict(testing_features)
```

Figure 4.27 – Source code for the optimized TPOT pipeline (Titanic)

And that does it for solving classification problems on the Titanic dataset in an automated fashion. You've now built two fully automated classification machine learning solutions. Let's wrap up this chapter next.

Summary

This was the second hands-on chapter in the book. You've learned how to solve classification machine learning tasks in an automated fashion with two in-depth examples on well-known datasets. Without any kind of doubt, you are now ready to use TPOT to solve any type of classification problem.

By now, you know how to solve regression and classification tasks. But what about parallel training? What about neural networks? The following chapter, *Chapter 5, Parallel Training with TPOT and Dask*, will teach you what parallel training is and how to utilize it with TPOT. Later, in *Chapter 6, Getting Started with Deep Learning – Crash Course in Neural Networks*, you'll reinforce your knowledge of basic deep learning and neural networks. As the icing on the cake, you'll learn how to use deep learning with TPOT in *Chapter 7, Neural Network Classifier with TPOT*.

Please feel encouraged to practice solving classification problems automatically with tools and techniques covered in this chapter.

Q&A

1. Can you explore the distribution of a categorical variable with bar charts? Explain.
2. Explain the confusion matrix and the terms true positive, true negative, false positive, and false negative.
3. What is precision? Explain by giving a practical example.
4. What is recall? Explain by giving a practical example.
5. What's the difference between accuracy and F1 score? When would you use F1 over accuracy?
6. What does "1" in the F1 score mean? Can this number be altered? What happens in that scenario?
7. During training, does TPOT output the value of your scoring metric for the train set or the test set? Explain.

5
Parallel Training with TPOT and Dask

In this chapter, you'll dive into a bit of a more advanced topic; that is, automated machine learning. You'll learn how to handle machine learning tasks in a parallel manner by distributing the work on a Dask cluster. This chapter will be more theoretical than the previous two, but you will still learn many useful things.

We'll cover essential topics and ideas behind parallelism in Python, and you'll learn how to achieve parallelism in a couple of different ways. Then, we'll dive deep into the Dask library, explore its basic functionality, and see how you can tie it with TPOT.

This chapter will cover the following topics:

- Introduction to parallelism in Python
- Introduction to the Dask library
- Training machine learning models with TPOT and Dask

Let's get started!

Technical requirements

No prior exposure to Dask or even parallel programming is required for you to read and understand this chapter. Previous experience is helpful, as fitting this big of a concept into a few pages is close to impossible. You should still be able to follow and fully understand everything written here as all of the concepts will be explained.

You can download the source code and dataset for this chapter here: `https://github.com/PacktPublishing/Machine-Learning-Automation-with-TPOT/tree/main/Chapter05`.

Introduction to parallelism in Python

Executing tasks sequentially (where the second one starts after the first one finishes) is required in some situations. For example, maybe the input of the second function relies on the output of the first one. If that's the case, these two functions (processes) can't be executed at the same time.

But more often than not, that's not the case. Just imagine your program is connecting to three different API endpoints before the dashboard is displayed. The first API returns the current weather conditions, the second one returns the stock prices, and the last one returns today's exchange rates. There's no point in making the API calls one after the other. They don't rely on each other, so running them sequentially would be a huge waste of time.

Not only that, but it would also be a waste of CPU cores. Most modern PCs have at least four CPU cores. If you're running things sequentially, you're only using a single core. Why not use all of them if you can?

One of the ways to achieve parallelism in Python is with multiprocessing. It is a process-based parallelism technique. As you would imagine, Python has a `multiprocessing` library built into it, and this section will teach you how to use it. With Python 3.2 and beyond, this library stopped being the recommended way of implementing multiprocessing in your apps. There's a new kid on the block, and its name is `concurrent.futures`. It's yet another built-in library you'll learn how to use in this section.

The simplest way to explain and understand multiprocessing is with Python's built-in `time` library. You can use it to track time differences and to pause program execution intentionally, among other things. This is just what we need because we can put in many print statements with some time gaps between them, and then see how the program acts when it's run sequentially and how it acts when it's run in parallel.

You will learn how multiprocessing works in Python through a couple of hands-on examples.

For starters, please take a look at the following code snippet. In it, the `sleep_func()` function has been declared. Its task is to print a message, pause the program executing for 1 second, and then to print another message as the function completes. We can monitor the time this function takes to run for an arbitrary number of times (let's say five) and then print out the execution time duration. The snippet is as follows:

```python
import time

def sleep_func():
    print('Sleeping for a 1 second')
    time.sleep(1)
    print('Done.')


if __name__ == '__main__':
    time_start = time.time()

    # Run the function 5 times
    sleep_func()
    sleep_func()
    sleep_func()
    sleep_func()
    sleep_func()

    time_stop = time.time()
    print(f'Took {round(time_stop - time_start, 2)} seconds to execute!')
```

The corresponding output is shown here:

```
Sleeping for a 1 second
Done.
Sleeping for a 1 second
Done.
Sleeping for a 1 second
Done.
Sleeping for a 1 second
Done.
Sleeping for a 1 second
```

```
Done.
Took 5.02 seconds to execute!
```

So, what happened here? Nothing unexpected, to say the least. The `sleep_func()` function executed sequentially five times. The execution time is approximately 5 seconds. You could also simplify the preceding snippet in the following manner:

```python
import time

def sleep_func():
    print('Sleeping for a 1 second')
    time.sleep(1)
    print('Done.')


if __name__ == '__main__':
    time_start = time.time()

    # Run the function 5 times in loop
    for _ in range(5):
        sleep_func()

    time_stop = time.time()
    print(f'Took {round(time_stop - time_start, 2)} seconds to
execute!')
```

The result is identical, as you would expect:

```
Sleeping for a 1 second
Done.
Sleeping for a 1 second
Done.
Sleeping for a 1 second
Done.
Sleeping for a 1 second
Done.
Sleeping for a 1 second
Done.
Took 5.01 seconds to execute!
```

Is there a problem with this approach? Well, yes. We're wasting both time and CPU cores. The functions aren't dependent in any way, so why don't we run them in parallel? As we mentioned previously, there are two ways of doing this. Let's examine the older way first, through the multiprocessing library.

It's a bit of a lengthy approach because it requires declaring a process, starting it, and joining it. It's not so tedious if you have only a few, but what if there are tens of processes in your program? It can become tedious fast.

The following code snippet demonstrates how to run the sleep_func() function three times in parallel:

```python
import time
from multiprocessing import Process

def sleep_func():
    print('Sleeping for a 1 second')
    time.sleep(1)
    print('Done.')


if __name__ == '__main__':
    time_start = time.time()

    process_1 = Process(target=sleep_func)
    process_2 = Process(target=sleep_func)
    process_3 = Process(target=sleep_func)

    process_1.start()
    process_2.start()
    process_3.start()

    process_1.join()
    process_2.join()
    process_3.join()

    time_stop = time.time()
    print(f'Took {round(time_stop - time_start, 2)} seconds to
execute!')
```

The output is shown here:

```
Sleeping for a 1 second
Sleeping for a 1 second
Sleeping for a 1 second
Done.
Done.
Done.
Took 1.07 seconds to execute!
```

As you can see, each of the three processes was launched independently and in parallel, so they all managed to finish in a single second.

Both `Process()` and `start()` are self-explanatory, but what is the `join()` function doing? Simply put, it tells Python to wait until the process is complete. If you call `join()` on all of the processes, the last two code lines won't execute until all of the processes are finished. For fun, try to remove the `join()` calls; you'll immediately get the gist.

You now have a basic intuition behind multiprocessing, but the story doesn't end here. Python 3.2 introduced a new, improved way of executing tasks in parallel. The `concurrent.futures` library is the best one available as of yet, and you'll learn how to use it next.

With it, you don't have to manage processes manually. Every executed function will return something, which is `None` in the case of our `sleep_func()` function. You can change it by returning the last statement instead of printing it. Furthermore, this new approach uses `ProcessPoolExecutor()` to run. You don't need to know anything about it; just remember that it is used to execute multiple processes at the same time. Codewise, simply put everything you want to run in parallel inside. This approach unlocks two new functions:

- `submit()`: Used to run the function in parallel. The returned results will be appended to a list so that we can print them (or do anything else) with the next function.

- `result()`: Used to obtain the returned value from the function. We'll simply print the result, but you're free to do anything else.

To recap, we'll append the results to a list, and then print them out as the functions finish executing. The following snippet shows you how to implement multiprocessing with the most recent Python approach:

```
import time
import concurrent.futures
```

```python
def sleep_func():
    print('Sleeping for a 1 second')
    time.sleep(1)
    return 'Done.'


if __name__ == '__main__':
    time_start = time.time()

    with concurrent.futures.ProcessPoolExecutor() as ppe:
        out = []
        for _ in range(5):
            out.append(ppe.submit(sleep_func))

        for curr in concurrent.futures.as_completed(out):
            print(curr.result())

    time_stop = time.time()
    print(f'Took {round(time_stop - time_start, 2)} seconds to
execute!')
```

The results are shown here:

```
Sleeping for a 1 second
Sleeping for a 1 second
Sleeping for a 1 second
Sleeping for a 1 second
Sleeping for a 1 second
Done.
Done.
Done.
Done.
Done.
Took 1.17 seconds to execute!
```

As you can see, the program behaves similarly to what we had previously, with a few added benefits – you don't have to manage processes on your own, and the syntax is much cleaner.

The one issue we have so far is the lack of function parameters. Currently, we're just calling a function that doesn't accept any parameters. That won't be the case most of the time, so it's important to learn how to handle function parameters as early as possible.

We'll introduce a single parameter to our `sleep_func()` function that allows us to specify how long the execution will be paused. The print statements inside the function are updated accordingly. The sleep times are defined within the `sleep_seconds` list, and the value is passed to `append()` at each iteration as a second parameter.

The entire snippet is shown here:

```python
import time
import concurrent.futures


def sleep_func(how_long: int):
    print(f'Sleeping for a {how_long} seconds')
    time.sleep(how_long)
    return f'Finished sleeping for {how_long} seconds.'


if __name__ == '__main__':
    time_start = time.time()
    sleep_seconds = [1, 2, 3, 1, 2, 3]

    with concurrent.futures.ProcessPoolExecutor() as ppe:
        out = []
        for sleep_second in sleep_seconds:
            out.append(ppe.submit(sleep_func, sleep_second))

        for curr in concurrent.futures.as_completed(out):
            print(curr.result())

    time_stop = time.time()
    print(f'Took {round(time_stop - time_start, 2)} seconds to
execute!')
```

The results are shown here:

```
Sleeping for 1 seconds
Sleeping for 2 seconds
Sleeping for 3 seconds
Sleeping for 1 seconds
Sleeping for 2 seconds
Sleeping for 3 seconds
Finished sleeping for 1 seconds.
Finished sleeping for 1 seconds.
Finished sleeping for 2 seconds.
```

```
Finished sleeping for 2 seconds.
Finished sleeping for 3 seconds.
Finished sleeping for 3 seconds.
Took 3.24 seconds to execute!
```

That's how you can handle function parameters in parallel processing. Keep in mind that the executing time won't be exactly the same on every machine, as the runtime duration will depend on your hardware. As a general rule, you should definitely see a speed improvement compared to a non-parallelized version of the script. You now know the basics of parallel processing. In the next section, you'll learn where Python's Dask library comes into the picture, and in the section afterward, you'll combine parallel programming, Dask, and TPOT in order to build machine learning models.

Introduction to the Dask library

You can think of Dask as one of the most revolutionary Python libraries for data processing at scale. If you are a regular pandas and NumPy user, you'll love Dask. The library allows you to work with data NumPy and pandas doesn't allow because they don't fit into the RAM.

Dask supports both NumPy array and pandas DataFrame data structures, so you'll quickly get up to speed with it. It can run either on your computer or a cluster, making it that much easier to scale. You only need to write the code once and then choose the environment that you'll run it in. It's that simple.

One other thing to note is that Dask allows you to run code in parallel with minimal changes. As you saw earlier, processing things in parallel means the execution time decreases, which is generally the behavior we want. Later, you'll learn how parallelism in Dask works with `dask.delayed`.

To get started, you'll have to install the library. Make sure the correct environment is activated. Then, execute the following from the Terminal:

```
pipenv install "dask[complete]"
```

There are other installation options. For example, you could install only the arrays or DataFrames module, but it's a good idea to install everything from the start. Don't forget to put quotes around the library name, as not doing so will result in an error.

If you've installed everything, you'll have access to three Dask collections – arrays, DataFrames, and bags. All of these can store datasets that are larger than your RAM size, and they can all partition data between RAM and a hard drive.

Let's start with Dask arrays and compare them with a NumPy alternative. You can create a NumPy array of ones with 1,000x1,000x1,000 dimensions by executing the following code cell in a Notebook environment. The `%%time` magic command is used to measure the time needed for the cell to finish with the execution:

```
%%time

import numpy as np

np_ones = np.ones((1000, 1000, 1000))
```

Constructing larger arrays than this one results in a memory error on my machine, but this will do just fine for the comparisons. The corresponding output is shown here:

```
CPU times: user 1.86 s, sys: 2.21 s, total: 4.07 s
Wall time: 4.35 s
```

As you can see, it took 4.35 seconds to create this array. Now, let's do the same with Dask:

```
%%time

import dask.array as da

da_ones = da.ones((1000, 1000, 1000))
```

As you can see, the only change is in the library import name. The executing time results will probably come as a surprise if this is your first encounter with the Dask library. They are shown here:

```
CPU times: user 677 µs, sys: 12 µs, total: 689 µs
Wall time: 696 µs
```

Yes, you are reading this right. Dask took 696 microseconds to create an array of identical dimensions, which is 6,250 times faster. Sure, you shouldn't expect this drastic reduction in execution time in the real world, but the differences should still be quite significant.

Next, let's take a look at Dask DataFrames. The syntax should, once again, feel very similar, so it shouldn't take you much time to learn the library. To fully demonstrate Dask's capabilities, we'll create some large datasets that won't be able to fit in the memory of a single laptop. To be more precise, we'll create 10 CSV files that are time series-based, each presenting data for a single year aggregated by seconds and measured through five different features. That's a lot, and it will definitely take some time to create, but you should end up with 10 datasets where each is around 1 GB in size. If you have a laptop with 8 GB of RAM like me, there's no way you could fit it in memory.

The following code snippet creates these datasets:

```
import pandas as pd
from datetime import datetime

for year in np.arange(2010, 2020):
    dates = pd.date_range(
        start=datetime(year=year, month=1, day=1),
        end=datetime(year=year, month=12, day=31),
        freq='S'
    )
    df = pd.DataFrame()
    df['Date'] = dates
    for i in range(5):
        df[f'X{i}'] = np.random.randint(low=0, high=100, size=len(df))

    df.to_csv(f'data/{year}.csv', index=False)

!ls data/
```

Just make sure to have this /data folder where your Notebook is and you'll be good to go. Also, make sure you have 10 GB of disk space if you're following along. The last line, !ls data/, lists all the files located in the data folder. Here's what you should see:

```
2010.csv 2012.csv 2014.csv 2016.csv 2018.csv
2011.csv 2013.csv 2015.csv 2017.csv 2019.csv
```

Now, let's take a look at how much time it takes pandas to read in a single CSV file and perform a simple aggregation. To be more precise, the dataset is grouped by month and the sum is extracted. The following code snippet demonstrates how to do this:

```
%%time

df = pd.read_csv('data/2010.csv', parse_dates=['Date'])
avg = df.groupby(by=df['Date'].dt.month).sum()
```

The results are shown here:

```
CPU times: user 26.5 s, sys: 9.7 s, total: 36.2 s
Wall time: 42 s
```

As you can see, it took pandas 42 seconds to perform this computation. Not too shabby, but what if you absolutely need to load in all of the datasets and perform computations? Let's explore that next.

You can use the `glob` library to get paths to desired files in a specified folder. You can then read all of them individually, and use the `concat()` function from pandas to stack them together. The aggregation is performed in the same way:

```
%%time

import glob
all_files = glob.glob('data/*.csv')
dfs = []

for fname in all_files:
    dfs.append(pd.read_csv(fname, parse_dates=['Date']))

df = pd.concat(dfs, axis=0)
agg = df.groupby(by=df['Date'].dt.year).sum()
```

There isn't much to say here – the Notebook simply breaks. Storing 10 GB+ of data into RAM isn't feasible for an 8 GB RAM machine. One way you could get around this would be to load data in chunks, but that's a headache of its own.

What can Dask do to help? Let's learn how to load in these CSVs with Dask and perform the same aggregation. You can use the following snippet to do so:

```
%%time

import dask.dataframe as dd

df = dd.read_csv('data/*.csv', parse_dates=['Date'])
agg = df.groupby(by=df['Date'].dt.year).sum().compute()
```

The results will once again surprise you:

```
CPU times: user 5min 3s, sys: 1min 11s, total: 6min 15s
Wall time: 3min 41s
```

That's correct – in less than 4 minutes, Dask managed to read over 10 GB of data to an 8 GB RAM machine. That alone should make you reconsider NumPy and pandas, especially if you're dealing with large amounts of data or you expect to deal with it in the near future.

Finally, there are Dask bags. They are used for storing and processing general Python data types that can't fit into memory – for example, log data. We won't explore this data structure, but it's nice to know it exists.

On the other hand, we will explore the concept of parallel processing with Dask. You learned in the previous section that there are no valid reasons to process data or perform any other operation sequentially, as the input of one doesn't rely on the output of another.

Dask delayed allows for parallel execution. Sure, you can still rely only on the multiprocessing concepts we learned earlier, but why? It can be a tedious approach, and Dask has something better to offer. With Dask, there's no need to change the programming syntax, as was the case with pure Python. You just need to annotate a function you want to be parallelized with the @dask.delayed decorator and you're good to go!

You can parallelize multiple functions and then place them inside a computational graph. That's what we'll do next.

The following code snippet declares two functions:

- cube(): Returns a cube of a number
- multiply(): Multiplies all numbers in a list and returns the product

Here are the library imports you'll need:

```
import time
import dask
import math
from dask import delayed, compute
```

Let's run the first function on five numbers and call the second function on the results to see what happens. Note the call to time.sleep() inside the cube() function. This will make spotting differences between parallelized and non-parallelized functions that much easier:

```
%%time

def cube(number: int) -> int:
    print(f'cube({number}) called!')
    time.sleep(1)
    return number ** 3

def multiply(items: list) -> int:
    print(f'multiply([{items}]) called!')
```

```
        return math.prod(items)

numbers = [1, 2, 3, 4, 5]
graph = multiply([cube(num) for num in numbers])
print(f'Total = {graph}')
```

This is your regular (sequential) data processing. There's nothing wrong with it, especially when there are so few and simple operations. The corresponding output is shown here:

```
cube(1) called!
cube(2) called!
cube(3) called!
cube(4) called!
cube(5) called!
multiply([[1, 8, 27, 64, 125]]) called!
Total = 1728000
CPU times: user 8.04 ms, sys: 4 ms, total: 12 ms
Wall time: 5.02 s
```

As expected, the code cell took around 5 seconds to run because of sequential execution. Now, let's see the modifications you have to make to parallelize these functions:

```
%%time

@delayed
def cube(number: int) -> int:
    print(f'cube({number}) called!')
    time.sleep(1)
    return number ** 3

@delayed
def multiply(items: list) -> int:
    print(f'multiply([{items}]) called!')
    return math.prod(items)

numbers = [1, 2, 3, 4, 5]
graph = multiply([cube(num) for num in numbers])
print(f'Total = {graph.compute()}')
```

So, there's only the `@delayed` decorator and a call to `compute()` on the graph. The results are displayed here:

```
cube(3) called!cube(2) called!cube(4) called!
cube(1) called!
cube(5) called!

multiply([[1, 8, 27, 64, 125]]) called!
Total = 1728000
CPU times: user 6.37 ms, sys: 5.4 ms, total: 11.8 ms
Wall time: 1.01 s
```

As expected, the whole thing took just over a second to run because of the parallel execution. The previously declared computational graph comes with one more handy feature – it's easy to visualize. You'll need to have *GraphViz* installed on your machine and as a Python library. The procedure is different for every OS, so we won't go through it here. A quick Google search will tell you how to install it. Once you're done, you can execute the following line of code:

```
graph.visualize()
```

The corresponding visualization is shown here:

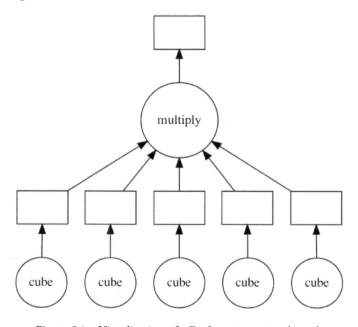

Figure 5.1 – Visualization of a Dask computational graph

As you can see from the graph, the `cube()` function is called five times in parallel, and its results are stored in the buckets above it. Then, the `multiply()` function is called with these values and stores the product in the top bucket.

That's all you need to know about the basics of Dask. You've learned how to work with Dask arrays and DataFrames, and also how to use Dask to process operations in parallel. Not only that, but you've also learned the crucial role Dask plays in modern-day data science and machine learning. Dataset sizes often exceed available memory, so modern solutions are required.

In the following section, you'll learn how to train TPOT automated machine learning models with Dask.

Training machine learning models with TPOT and Dask

Optimizing machine learning pipelines is, before everything, a time-consuming process. We can shorten it potentially significantly by running things in parallel. Dask and TPOT work great when combined, and this section will teach you how to train TPOT models on a Dask cluster. Don't let the word "cluster" scare you, as your laptop or PC will be enough.

You'll have to install one more library to continue, and it is called `dask-ml`. As its name suggests, it's used to perform machine learning with Dask. Execute the following from the Terminal to install it:

```
pipenv install dask-ml
```

Once that's done, you can open up Jupyter Lab or your favorite Python code editor and start coding. Let's get started:

1. Let's start with library imports. We'll also make a dataset decision here. This time, we won't spend any time on data cleaning, preparation, or examination. The goal is to have a dataset ready as soon as possible. The `load_digits()` function from scikit-learn comes in handy because it is designed to fetch many 8x8 pixel digit images for classification.

   As some of the libraries often fill up your screen with unnecessary warnings, we'll use the `warnings` library to ignore them. Refer to the following snippet for all the library imports:

   ```
   import tpot
   from tpot import TPOTClassifier
   from sklearn.datasets import load_digits
   ```

```
from sklearn.model_selection import train_test_split
from dask.distributed import Client

import warnings
warnings.filterwarnings('ignore')
```

The only new thing here is the `Client` class from `dask.distributed`. It is used to establish a connection with the Dask cluster (your computer, in this case).

2. You'll now make an instance of the client. This will immediately start the Dask cluster and use all the CPU cores you have available. Here's the code for instance creation and checking where the cluster runs:

```
client = Client()
client
```

Once executed, you should see the following output:

Client

Scheduler: tcp://127.0.0.1:59459
Dashboard: http://127.0.0.1:8787/status

Cluster

Workers: 4
Cores: 8
Memory: 8.59 GB

Figure 5.2 – Information on the Dask cluster

You can click on the dashboard link, and it will take you to `http://127.0.0.1:8787/status`. The following screenshot shows what the dashboard should look like when it's opened for the first time (no tasks running):

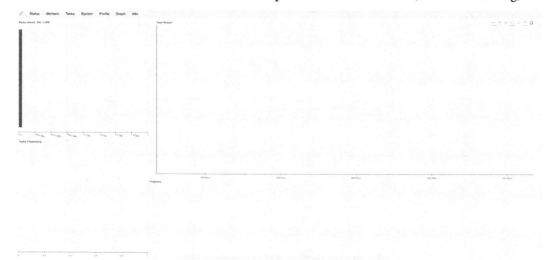

Figure 5.3 – Dask cluster dashboard (no running tasks)

The dashboard will become much more colorful once you start training the models. We'll do the necessary preparation next.

3. You can call the `load_digits()` function to get the image data and then use the `train_test_split()` function to split the images into subsets for training and testing. The train/test ratio is 50:50 for this example, as we don't want to spend too much time on the training. The ratio should be higher for the training set in almost any scenario, so make sure to remember that.

    Once the split is done, you can call `.shape` on the subsets to check their dimensionality. Here's the entire code snippet:

```
digits = load_digits()

X_train, X_test, y_train, y_test = train_test_split(
    digits.data,
    digits.target,
    test_size=0.5,
)

X_train.shape, X_test.shape
```

The corresponding output is shown in the following figure:

```
((898, 64), (899, 64))
```

Figure 5.4 – Dimensionality of training and testing subsets

Next stop – model training.

4. You now have everything needed to train models with TPOT and Dask. You can do so in a very similar fashion to what you did previously. The key parameter here is `use_dask`. You should set it to `True` if you want to use Dask for training. The other parameters are well known:

```
estimator = TPOTClassifier(
    n_jobs=-1,
    random_state=42,
    use_dask=True,
    verbosity=2,
    max_time_mins=10
)
```

Now, you're ready to call the `fit()` function and train the model on the training subset. Here's a line of code for doing so:

```
estimator.fit(X_train, y_train)
```

The appearance of the Dask dashboard will change immediately after you start training the model. Here's what it will look like a couple of minutes into the process:

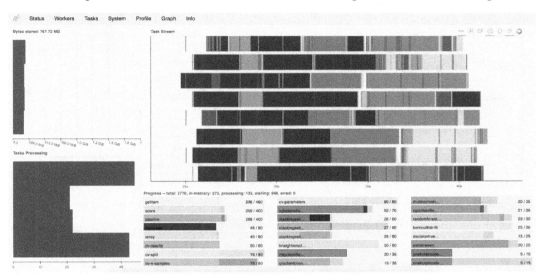

Figure 5.5 – Dask dashboard during training

After 10 minutes, TPOT will finish optimizing the pipeline, and you'll see the following output in your Notebook:

```
HBox(children=(FloatProgress(value=0.0, description='Optimization Progress', style=ProgressStyle(description_w…

Generation 1 - Current best internal CV score: 0.9855121042830539

Generation 2 - Current best internal CV score: 0.9855183116076971

Generation 3 - Current best internal CV score: 0.9866294227188082

10.38 minutes have elapsed. TPOT will close down.
TPOT closed during evaluation in one generation.
WARNING: TPOT may not provide a good pipeline if TPOT is stopped/interrupted in a early generation.

TPOT closed prematurely. Will use the current best pipeline.

Best pipeline: KNeighborsClassifier(GaussianNB(LinearSVC(input_matrix, C=0.1, dual=False, loss=squared_hinge, penalty=l1, tol=0.01)), n_neighb
ors=2, p=2, weights=distance)
TPOTClassifier(max_time_mins=10, n_jobs=-1, random_state=42, use_dask=True,
               verbosity=2)
```

Figure 5.6 – TPOT optimization outputs

And that's all you need to do to combine TPOT and Dask.

You now know how to train models on a Dask cluster, which is the recommended way of doing things for larger datasets and more challenging problems.

Summary

This chapter was packed with information not only about TPOT and training models in a parallel manner, but also about parallelism in general. You've learned a lot – from how to parallelize basic functions that do nothing but sleep for a while, to parallelizing function with parameters and Dask fundamentals, to training machine learning models with TPOT and Dask on a Dask cluster.

By now, you know how to solve regression and classification tasks in an automated manner, and how to parallelize the training process. The following chapter, *Chapter 6, Getting Started with Deep Learning – Crash Course in Neural Networks*, will provide you with the required knowledge on neural networks. It will form a basis for *Chapter 7, Neural Network Classifier with TPOT*, where we'll dive deep into training automated machine learning models with state-of-the-art neural network algorithms.

As always, please feel free to practice solving both regression and classification tasks with TPOT, but this time, try to parallelize the process with Dask.

Q&A

1. Define the term "parallelism."
2. Explain which types of tasks can and can't be parallelized.
3. List and explain three options for implementing parallelism in your applications (all are listed in this chapter).
4. What is Dask and what makes it superior to NumPy and pandas for larger datasets?
5. Name and explain three basic data structures that are implemented in Dask.
6. What is a Dask cluster?
7. What do you have to do to tell TPOT it should use Dask for training?

Section 3: Advanced Examples and Neural Networks in TPOT

This section is aimed toward more advanced users. Topics such as neural networks are introduced, and neural network classifiers are built in TPOT. Finally, a predictive model is deployed as a REST API both locally and to the cloud and is then used to make predictions in real time.

This section contains the following chapters:

- *Chapter 6, Getting Started with Deep Learning – Crash Course in Neural Networks*
- *Chapter 7, Neural Network Classifier with TPOT*
- *Chapter 8, TPOT Model Deployment*
- *Chapter 9, Using the Deployed TPOT Model in Production*

6

Getting Started with Deep Learning: Crash Course in Neural Networks

In this chapter, you'll learn the basics of deep learning and artificial neural networks. You'll discover the basic idea and theory behind these topics and how to train simple neural network models with Python. The chapter will serve as an excellent primer for the upcoming chapters, where the ideas of pipeline optimization and neural networks are combined.

We'll cover the essential topics and ideas behind deep learning, why it has gained popularity in the last few years, and the cases in which neural networks work better than traditional machine learning algorithms. You'll also get hands-on experience in coding your own neural networks, both from scratch and through pre-made libraries.

This chapter will cover the following topics:

- An overview of deep learning
- Introducing artificial neural networks
- Using neural networks to classify handwritten digits
- Comparing neural networks in regression and classification

Technical requirements

No prior experience with deep learning and neural networks is necessary. You should be able to understand the basics from this chapter alone. Previous experience is helpful, as deep learning isn't something you can learn in one sitting.

You can download the source code and dataset for this chapter here: `https://github.com/PacktPublishing/Machine-Learning-Automation-with-TPOT/tree/main/Chapter06`.

Overview of deep learning

Deep learning is a subfield of machine learning that focuses on neural networks. Neural networks aren't that new as a concept – they were introduced back in the 1940s but didn't gain much in popularity until they started winning data science competitions (somewhere around 2010).

Potentially the biggest year for deep learning and AI was 2016, all due to a single event. *AlphaGo*, a computer program that plays the board game Go, defeated the highest-ranking player in the world. Before this event, Go was considered to be a game that computers couldn't master, as there are so many potential board configurations.

As mentioned before, deep learning is based on neural networks. You can think of neural networks as **directed acyclic graphs** – a graph consisting of vertices (nodes) and edges (connections). The input layer (the first layer, on the far left side) takes in the raw data from your datasets, passes it through one or multiple hidden layers, and constructs an output.

You can see an example architecture of a neural network in the following diagram:

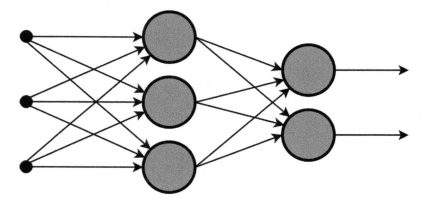

Figure 6.1 – An example neural network architecture

The small black nodes on the far left side represent the input data – data that comes directly from your dataset. These values are then connected with the hidden layers, with their respective weights and biases. A common way to refer to these weights and biases is by using the term **tunable parameters**. We'll address this term and show how to calculate them in the next section.

Every node of a neural network is called a **neuron**. Let's take a look at the architecture of an individual neuron:

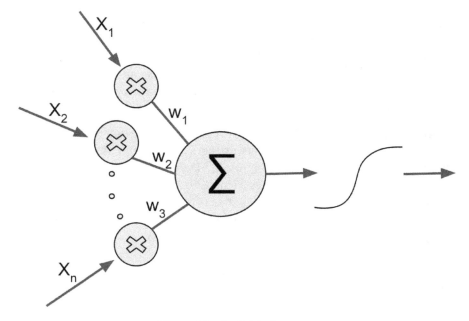

Figure 6.2 – Individual neuron

The X's correspond to the values either from the input layer or from the previous hidden layer. These values are multiplied together ($x1 * w1$, $x2 * w2$) and then added together ($x1w1 + x2w2$). After the summation, a bias term is added, and finally, everything is passed through an **activation function**. This function determines if the neuron will "fire" or not. It's something like an on-off switch, in the simplest terms.

A brief explanation of weights and biases is shown here:

- Weights:

  a) Multiplied with values from the previous layer

  b) Can change the magnitude or entirely flip the value from positive to negative

  c) In function terms – adjusting the weight changes the slope of the function

- Biases:

  a) Interpreted as an offset of the function

  b) An increase in bias leads to an upward shift of a function

  c) A decrease in bias leads to a downward shift of a function

There are many types of neural network architectures besides artificial neural networks, and they are discussed here:

- **Convolutional neural networks (CNNs)** – a type of neural network most commonly applied to analyzing images. They are based on the convolution operation – an operation between two matrices in which the second one slides (convolves) over the first one and computes element-wise multiplication. The goal of this operation is to find a sliding matrix (kernel) that can extract the correct features from the input image and hence make image classification tasks easy.

- **Recurrent neural networks (RNNs)** – a type of neural network most commonly used on sequence data. Today these networks are applied in many tasks, such as handwriting recognition, speech recognition, machine translation, and time series forecasting. The RNN model processes a single element in the sequence at a time. After processing, the new updated unit's state is passed down to the next time step. Imagine predicting a single character based on the previous n characters; that's the general gist.

- **Generative adversarial networks (GANs)** – a type of neural network most commonly used to create new samples after learning from real data. The GAN architecture comprises two separate models – generators and discriminators. The job of a generator model is to make fake images and send them to the discriminator. The discriminator works like a judge and tries to tell whether an image is fake or not.

- **Autoencoders** – unsupervised learning techniques, designed to learn a low-dimensional representation of a high-dimensional dataset. In a way, they work similarly to **Principal Component Analysis (PCA)**.

These four deep learning concepts won't be covered in this book. We'll focus only on artificial neural networks, but it's good to know they exist in case you want to dive deeper on your own.

The next section looks at artificial neural networks and shows you how to implement them in Python, both from scratch and with data science libraries.

Introducing artificial neural networks

The fundamental building block of an artificial neural network is the neuron. By itself, a single neuron is useless, but it can have strong predictive power when combined into a more complex network.

If you can't reason why, think about your brain and how it works for a minute. Just like artificial neural networks, it is also made from millions of neurons, which function only when there's communication between them. Since artificial neural networks try to imitate the human brain, they need to somehow replicate neurons in the brain and connections between them (weights). This association will be made less abstract throughout this section.

Today, artificial neural networks can be used to tackle any problem that regular machine learning algorithms can. In a nutshell, if you can solve a problem with linear or logistic regression, you can solve it with neural networks.

Before we can explore the complexity and inner workings of an entire network, we have to start simple – with the theory of a single neuron.

Theory of a single neuron

Modeling a single neuron is easy with Python. For example, let's say a neuron receives values from five other neurons (inputs, or X's). Let's examine this behavior visually before implementing it in code. The following diagram shows how a single neuron looks when receiving values from five neurons in the previous layers (we're modeling the neuron on the right):

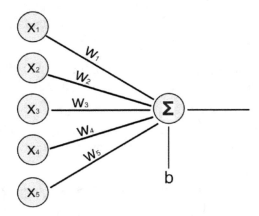

Figure 6.3 – Modeling a single neuron

The X's represent input features, either from the raw data or from the previous hidden layer. Each input feature has a weight assigned to it, denoted with W's. Corresponding input values and weights are multiplied and summed, and then the bias term (b) is added on top of the result.

The formula for calculating the output value of our neuron is as follows:

$$output = x_1 w_1 + x_2 w_2 + x_3 w_3 + x_4 w_4 + x_5 w_5 + b$$

Let's work with concrete values to get this concept a bit clearer. The following diagram looks identical to Figure 6.3, but has actual numbers instead of variables:

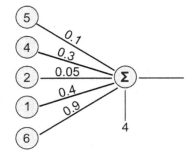

Figure 6.4 – Neuron value calculation

We can plug the values directly into the preceding formula to calculate the value:

$$output = x_1 w_1 + x_2 w_2 + x_3 w_3 + x_4 w_4 + x_5 w_5 + b$$

$$output = (5 \times 0.1) + (4 \times 0.3) + (2 \times 0.05) + (1 \times 0.4) + (6 \times 0.9) + 4$$

$$output = 0.5 + 1.2 + 0.1 + 0.4 + 5.4 + 4$$

$$output = 11.6$$

In reality, single neurons get their value from potentially thousands of neurons in the previous layers, so calculating values manually and expressing visually isn't practical.

Even if you decide to do so, that's only a single forward pass. Neural networks learn during the backward pass, which is much more complicated to calculate by hand.

Coding a single neuron

Next, let's see how you can semi-automate neuron value calculation with Python:

1. To start, let's declare input values, their respective weights, and a value for the bias term. The first two are lists, and the bias is just a number:

    ```
    inputs = [5, 4, 2, 1, 6]
    weights = [0.1, 0.3, 0.05, 0.4, 0.9]
    bias = 4
    ```

    That's all you need to calculate the output value. Let's examine what your options are next.

2. There are three simple methods for calculating neuron output values. The first one is the most manual, and that is to explicitly multiply corresponding inputs and weights, adding them together with the bias.

    Here's a Python implementation:

    ```
    output = (inputs[0] * weights[0] +
              inputs[1] * weights[1] +
              inputs[2] * weights[2] +
              inputs[3] * weights[3] +
              inputs[4] * weights[4] +
              bias)
    output
    ```

You should see a value of 11.6 printed out after executing this code. To be more precise, the value should be 11.600000000000001, but don't worry about this calculation error.

The next method is a bit more scalable, and it boils down to iterating through inputs and weights at the same time and incrementing the variable declared earlier for the output. After the loop finishes, the bias term is added. Here's how to implement this calculation method:

```
output = 0
for x, w in zip(inputs, weights):
    output += x * w
output += bias
output
```

The output is still identical, but you can immediately see how much more scalable this option is. Just imagine using the first option if the previous network layer had 1,000 neurons – it's not even remotely convenient.

The third and preferred method is to use a scientific computing library, such as NumPy. With it, you can calculate the vector dot product and add the bias term. Here's how:

```
import numpy as np

output = np.dot(inputs, weights) + bias
output
```

This option is the fastest, both to write and to execute, so it's the preferred one.

You now know how to code a single neuron – but neural networks employ layers of neurons. You'll learn more about layers next.

Theory of a single layer

To make things simpler, think of layers as vectors or simple groups. Layers aren't some complicated or abstract data structure. In code terms, you can think of them as lists. They contain a number of neurons.

Coding a single layer of neurons is quite similar to coding a single neuron. We still have the same inputs, as they are coming either from a previous hidden layer or an input layer. What changes are weights and biases. In code terms, weights aren't treated as a list anymore, but as a list of lists instead (or a matrix). Similarly, bias is now a list instead of a scalar value.

Put simply, your matrix of weights will have as many rows as there are neurons in the new layer and as many columns as there are neurons in the previous layer. Let's take a look at a sample diagram to make this concept a bit less abstract:

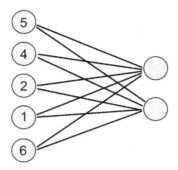

Figure 6.5 – Layer of neurons

Weight values deliberately weren't placed on the previous diagram, as it would look messy. To implement this layer in code, you'll need to have the following structures:

- Vector of inputs (1 row, 5 columns)
- Matrix of weights (2 rows, 5 columns)
- Vector of biases (1 row, 2 columns)

A matrix multiplication rule from linear algebra states that two matrices need to be of shapes (m, n) and (n, p) in order to produce an (m, p) matrix after multiplication. Bearing that in mind, you could easily perform matrix multiplication by transposing the matrix of weights.

Mathematically, here's the formula you can use to calculate the values of the output layer:

$$output = x \cdot W^T + b$$

Here, the following applies:

- x is the vector of inputs.
- W is the matrix of weights.
- b is the vector of biases.

Let's declare values for all of these and see how to calculate the values for an output layer:

$$x = \begin{bmatrix} 5 \\ 4 \\ 2 \\ 1 \\ 6 \end{bmatrix}, \quad W = \begin{bmatrix} 0.1 & 0.3 & 0.05 & 0.4 & 0.9 \\ 0.3 & 0.15 & 0.4 & 0.7 & 0.2 \end{bmatrix}, \quad b = \begin{bmatrix} 4 \\ 2 \end{bmatrix}$$

The previously mentioned formula can now be used to calculate the values of the output layer:

$$output = x \cdot W^T + b$$

$$output = \begin{bmatrix} 5 \\ 4 \\ 2 \\ 1 \\ 6 \end{bmatrix} \cdot \begin{bmatrix} 0.1 & 0.3 & 0.05 & 0.4 & 0.9 \\ 0.3 & 0.15 & 0.4 & 0.7 & 0.2 \end{bmatrix}^T + \begin{bmatrix} 4 \\ 2 \end{bmatrix}$$

$$output = \begin{bmatrix} 5 \\ 4 \\ 2 \\ 1 \\ 6 \end{bmatrix} \cdot \begin{bmatrix} 0.1 & 0.3 \\ 0.3 & 0.15 \\ 0.05 & 0.4 \\ 0.4 & 0.7 \\ 0.9 & 0.2 \end{bmatrix} + \begin{bmatrix} 4 \\ 2 \end{bmatrix}$$

$$output = \begin{bmatrix} 5 \times 0.1 + 4 \times 0.3 + 2 \times 0.05 + 1 \times 0.4 + 6 \times 0.9 \\ 5 \times 0.3 + 4 \times 0.15 + 2 \times 0.4 + 1 \times 0.7 + 6 \times 0.2 \end{bmatrix} + \begin{bmatrix} 4 \\ 2 \end{bmatrix}$$

$$output = \begin{bmatrix} 7.6 \\ 4.8 \end{bmatrix} + \begin{bmatrix} 4 \\ 2 \end{bmatrix}$$

$$output = \begin{bmatrix} 11.6 \\ 6.8 \end{bmatrix}$$

And that's essentially how you can calculate outputs for an entire layer. The calculations will grow in size for actual neural networks, as there are thousands of neurons per layer, but the logic behind the math is identical.

You can see how tedious it is to calculate layer outputs manually. You'll learn how to calculate the values in Python next.

Coding a single layer

Let's now examine three ways in which you could calculate the output values for a single layer. As with single neurons, we'll start with the manual approach and finish with a NumPy one-liner.

You'll have to declare values for inputs, weights, and biases first, so here's how to do that:

```
inputs = [5, 4, 2, 1, 6]
weights = [
     [0.1, 0.3, 0.05, 0.4, 0.9],
     [0.3, 0.15, 0.4, 0.7, 0.2]
]
biases = [4, 2]
```

Let's proceed with calculating the values of the output layer:

1. Let's start with the manual approach. No, we won't do the same procedure as with neurons. You could, of course, but it would look too messy and impractical. Instead, we'll immediately use the `zip()` function to iterate over the `weights` matrix and `biases` array and calculate the value of a single output neuron.

    This procedure is repeated for however many neurons there are, and each output neuron is appended to a list that represents the output layer.

    Here's the entire code snippet:

    ```
    layer = []

    for n_w, n_b in zip(weights, biases):
        output = 0
        for x, w in zip(inputs, n_w):
            output += x * w
        output += n_b
        layer.append(output)

    layer
    ```

    The result is a list with the value `[11.6, 6.8]`, which are the same results we got from the manual calculation earlier.

    While this approach works, it's still not optimal. Let's see how to improve next.

2. You'll now calculate the values of the output layer by taking the vector dot product between input values and every row of the `weights` matrix. The bias term will be added after this operation is completed.

    Let's see how it works in action:

    ```
    import numpy as np

    layer = []

    for n_w, n_b in zip(weights, biases):
        layer.append(np.dot(inputs, n_w) + n_b)

    layer
    ```

    The layer values are still identical – `[11.6, 6.8]`, and this approach is a bit more scalable than the previous one. It can still be improved upon. Let's see how next.

3. You can perform a matrix multiplication between inputs and transposed weights and add the corresponding biases with a single line of Python code. Here's how:

    ```
    layer = np.dot(inputs, np.transpose(weights)) + biases
    layer
    ```

    That's the recommended way if, for some reason, you want to calculate outputs manually. NumPy handles it completely, so it's the fastest one at the same time.

You now know how to calculate output values both for a single neuron and for a single layer of a neural network. So far, we haven't covered a crucial idea in neural networks that decides whether a neuron will "fire" or "activate" or not. These are called activation functions, and we'll cover them next.

Activation functions

Activation functions are essential for the output of neural networks, and hence to the output of the deep learning model. They are nothing but mathematical equations, and relatively simple ones to be precise. Activation functions are those that determine whether the neuron should be "activated" or not.

Another way to think about the activation function is as a sort of gate that stands between the input coming into the current neuron and its output, which goes to the next layer. Activation function can be as simple as a step function (turns neurons on or off), or a bit more complicated and non-linear. It's the non-linear functions that prove useful in learning complex data and providing accurate predictions.

We'll go over a couple of the most common activation functions next.

Step function

The step function is based on a threshold. If the value coming in is above the threshold, the neuron is activated. That's why we can say the step function serves as an on-off switch – there are no values in between.

You can easily use Python and NumPy to declare and visualize a basic step function. The procedure is shown here:

1. To start, you'll have to define a step function. The typical threshold value is 0, so the neuron will activate if and only if the value passed in to the function is greater than 0 (the input value being the sum of the previous inputs multiplied by the weights and added bias).

    This kind of logic is trivial to implement in Python:

    ```python
    def step_function(x):
        return 1 if x > 0 else 0
    ```

2. You can now declare a list of values that will serve as an input to this function, and then apply `step_function()` to this list. Here's an example:

    ```python
    xs = np.arange(-10, 10, step=0.1)
    ys = [step_function(x) for x in xs]
    ```

3. Finally, you can visualize the function with the help of the Matplotlib library in just two lines of code:

    ```python
    plt.plot(xs, ys, color='#000000', lw=3)
    plt.title('Step activation function', fontsize=20)
    ```

You can see how the function works visually in the following diagram:

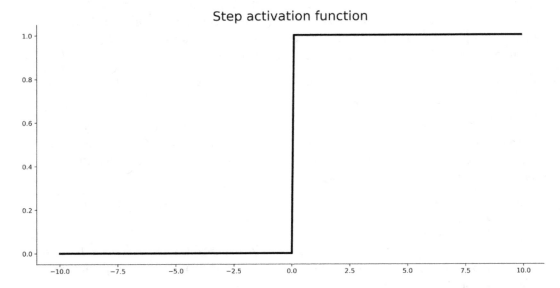

Figure 6.6 – Step activation function

The biggest problem of the step function is that it doesn't allow multiple outputs – only two. We'll dive into a set of non-linear functions next, and you'll see what makes them different.

Sigmoid function

The sigmoid activation function is frequently referred to as the logistic function. It is a very popular function in the realm of neural networks and deep learning. It essentially transforms the input into a value between 0 and 1.

You'll see how the function works later, and you'll immediately notice an advantage over the step function – the gradient is smooth, so there are no jumps in the output values. For example, you wouldn't get a jump from 0 to 1 if the value changed slightly (for example, from -0.000001 to 0.0000001).

The sigmoid function does suffer from a common problem in deep learning – **vanishing gradient**. It is a problem that often occurs during backpropagation (a process of learning in neural networks, way beyond this chapter's scope). Put simply, the gradient "vanishes" during the backward pass, making it impossible for the network to learn (tweak weights and biases), as the suggested tweaks are too close to zero.

You can use Python and NumPy to easily declare and visualize the sigmoid function. The procedure is shown here:

1. To start, you'll have to define the sigmoid function. Its formula is pretty well established: *(1 / (1 + exp(-x)))*, where *x* is the input value.

    Here's how to implement this formula in Python:

    ```python
    def sigmoid_function(x):
        return 1 / (1 + np.exp(-x))
    ```

2. You can now declare a list of values that will serve as an input to this function, and then apply `sigmoid_function()` to this list. Here's an example:

    ```python
    xs = np.arange(-10, 10, step=0.1)
    ys = [step_function(x) for x in xs]
    ```

3. Finally, you can visualize the function with the help of the Matplotlib library in just two lines of code:

    ```python
    plt.plot(xs, ys, color='#000000', lw=3)
    plt.title(Sigmoid activation function', fontsize=20)
    ```

    You can see how the function works visually in the following diagram:

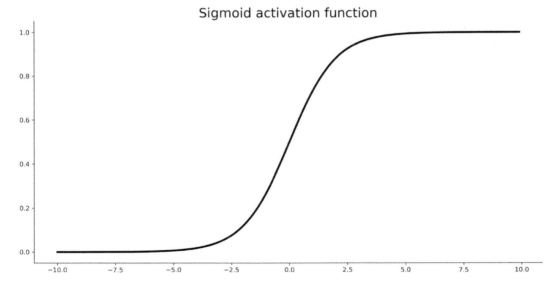

Figure 6.7 – Sigmoid activation function

One big disadvantage is that the values returned by the sigmoid function are not centered around zero. This is a problem because modeling inputs that are highly negative or highly positive gets harder. The hyperbolic tangent function fixes this problem.

Hyperbolic tangent function

Hyperbolic tangent function (or TanH) is closely related to the sigmoid function. It's also a type of activation function that suffers from the vanishing gradient issue, but its outputs are centered around zero – as the function ranges from -1 to +1.

This makes it much easier to model inputs that are highly negative or highly positive. You can use Python and NumPy to easily declare and visualize the hyperbolic tangent function. The procedure is shown here:

1. To start, you'll have to define the hyperbolic tangent function. You can use the `tanh()` function from NumPy for the implementation.

    Here's how to implement it in Python:
    ```
    def tanh_function(x):
        return np.tanh(x)
    ```

2. You can now declare a list of values that will serve as an input to this function, and then apply the `tanh_function()` to this list. Here's an example:
    ```
    xs = np.arange(-10, 10, step=0.1)
    ys = [step_function(x) for x in xs]
    ```

3. Finally, you can visualize the function with the help of the Matplotlib library in just two lines of code:
    ```
    plt.plot(xs, ys, color='#000000', lw=3)
    plt.title(Tanh activation function', fontsize=20)
    ```

You can see how the function works visually in the following diagram:

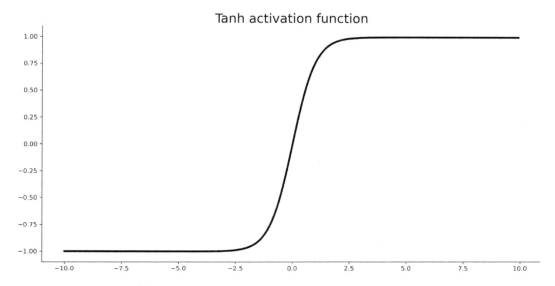

Figure 6.8 – Hyperbolic tangent activation function

To train and optimize neural networks efficiently, you need an activation function that acts as a linear function but is non-linear in nature, allowing the network to learn the complex relationships in the data. That's where the last activation function in this section comes in.

Rectified linear unit function

The rectified linear unit (or ReLU) function is an activation function you can see in most modern-day deep learning architectures. Put simply, it returns the larger of two values between 0 and x, where x is the input value.

ReLU is one of the most computationally efficient functions, and it allows the relatively quick finding of the convergence point. You'll see how to implement it in Python next:

1. To start, you'll have to define the ReLU function. This can be done entirely from scratch or with NumPy, as you only have to find the larger values of the two (0 and x).

    Here's how to implement ReLU in Python:

    ```python
    def relu_function(x):
        return np.maximum(0, x)
    ```

2. You can now declare a list of values that will serve as an input to this function, and then apply `relu_function()` to this list. Here's an example:

```
xs = np.arange(-10, 10, step=0.1)
ys = [step_function(x) for x in xs]
```

3. Finally, you can visualize the function with the help of the Matplotlib library in just two lines of code:

```
plt.plot(xs, ys, color='#000000', lw=3)
plt.title(ReLU activation function', fontsize=20)
```

You can see how the function works visually in the following diagram:

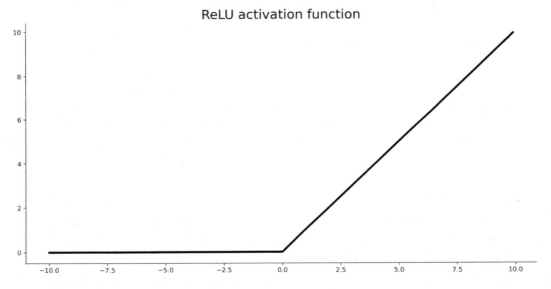

Figure 6.9 – ReLU activation function

And that's ReLU in a nutshell. You can use the default version or any of the variations (for example, leaky ReLU or parametric ReLU), depending on the use case.

You now know enough of the theory to code a basic neural network with Python. We haven't covered all of the theoretical topics, so terms such as loss, gradient descent, backpropagation, and others may still feel abstract. We'll try to demystify them in the hands-on example that's coming up next.

Using neural networks to classify handwritten digits

The "hello world" of deep learning is training a model that can classify handwritten digits. That's just what you'll do in this section. It will only require a couple of lines of code to implement with the TensorFlow library.

Before you can proceed, you'll have to install TensorFlow. The process is a bit different depending on whether you're on Windows, macOS, or Linux, and whether you have a CUDA-compatible GPU or not. You can refer to the official installation instructions: `https://www.tensorflow.org/install`. The rest of this section assumes you have TensorFlow 2.x installed. Here are the steps to follow:

1. To start, you'll have to import the TensorFlow library along with some additional modules. The `datasets` module enables you to download data straight from the notebook. The `layers` and `models` modules will be used later to design the architecture of the neural network.

    Here's the code snippet for the imports:

    ```
    import tensorflow as tf
    from tensorflow.keras import datasets, layers, models
    ```

2. You can now proceed with data gathering and preparation. A call to `datasets.mnist.load_data()` will download train and test images alongside the train and test labels. The images are grayscale and 28x28 pixels in size. This means you'll have a bunch of 28x28 matrices with values ranging from 0 (black) to 255 (white).

    You can then further prepare the dataset by rescaling the images – dividing the values by 255 to bring everything into a zero-to-one range:

    ```
    (train_images, train_labels), (test_images, test_labels)
    = datasets.mnist.load_data()
    train_images, test_images = train_images / 255.0, test_
    images / 255.0
    ```

    Here's what you should see in your notebook:

```
Downloading data from https://storage.googleapis.com/tensorflow/tf-keras-datasets/mnist.npz
11493376/11490434 [==============================] - 1s 0us/step
```

Figure 6.10 – Downloading the MNIST dataset

3. Furthermore, you can inspect the matrix values for one of the images to see if you can spot the pattern inside.

    The following line of code makes it easy to inspect matrices – it prints them and rounds all floating-point numbers to a single decimal point:

    ```
    print('\n'.join([''.join(['{:4}'.format(round(item, 1))
    for item in row]) for row in train_images[0]]))
    ```

    The results are shown in the following screenshot:

```
0.0 0.0 0.0 0.0 0.0 0.0 0.0 0.0 0.0 0.0 0.0 0.0 0.0 0.0 0.0 0.0 0.0 0.0 0.0 0.0 0.0 0.0 0.0 0.0 0.0 0.0 0.0 0.0
0.0 0.0 0.0 0.0 0.0 0.0 0.0 0.0 0.0 0.0 0.0 0.0 0.0 0.0 0.0 0.0 0.0 0.0 0.0 0.0 0.0 0.0 0.0 0.0 0.0 0.0 0.0 0.0
0.0 0.0 0.0 0.0 0.0 0.0 0.0 0.0 0.0 0.0 0.0 0.0 0.0 0.0 0.0 0.0 0.0 0.0 0.0 0.0 0.0 0.0 0.0 0.0 0.0 0.0 0.0 0.0
0.0 0.0 0.0 0.0 0.0 0.0 0.0 0.0 0.0 0.0 0.0 0.0 0.0 0.0 0.0 0.0 0.0 0.0 0.0 0.0 0.0 0.0 0.0 0.0 0.0 0.0 0.0 0.0
0.0 0.0 0.0 0.0 0.0 0.0 0.0 0.0 0.0 0.0 0.0 0.0 0.0 0.0 0.0 0.0 0.0 0.0 0.0 0.0 0.0 0.0 0.0 0.0 0.0 0.0 0.0 0.0
0.0 0.0 0.0 0.0 0.0 0.0 0.0 0.0 0.0 0.0 0.0 0.0 0.1 0.1 0.1 0.5 0.5 0.7 0.1 0.7 1.0 1.0 0.5 0.0 0.0 0.0 0.0 0.0
0.0 0.0 0.0 0.0 0.0 0.0 0.0 0.0 0.0 0.1 0.1 0.4 0.6 0.7 1.0 1.0 1.0 1.0 1.0 0.9 0.7 1.0 0.9 0.8 0.3 0.0 0.0 0.0
0.0 0.0 0.0 0.0 0.0 0.0 0.0 0.0 0.2 0.9 1.0 1.0 1.0 1.0 1.0 1.0 1.0 1.0 0.4 0.3 0.3 0.2 0.2 0.0 0.0 0.0 0.0 0.0
0.0 0.0 0.0 0.0 0.0 0.0 0.0 0.0 0.1 0.9 1.0 1.0 1.0 1.0 0.8 0.7 1.0 0.9 0.0 0.0 0.0 0.0 0.0 0.0 0.0 0.0 0.0 0.0
0.0 0.0 0.0 0.0 0.0 0.0 0.0 0.0 0.0 0.3 0.6 0.4 1.0 1.0 0.8 0.0 0.0 0.2 0.6 0.0 0.0 0.0 0.0 0.0 0.0 0.0 0.0 0.0
0.0 0.0 0.0 0.0 0.0 0.0 0.0 0.0 0.0 0.0 0.1 0.0 0.6 1.0 0.4 0.0 0.0 0.0 0.0 0.0 0.0 0.0 0.0 0.0 0.0 0.0 0.0 0.0
0.0 0.0 0.0 0.0 0.0 0.0 0.0 0.0 0.0 0.0 0.0 0.0 0.5 1.0 0.7 0.0 0.0 0.0 0.0 0.0 0.0 0.0 0.0 0.0 0.0 0.0 0.0 0.0
0.0 0.0 0.0 0.0 0.0 0.0 0.0 0.0 0.0 0.0 0.0 0.0 0.0 0.7 1.0 0.3 0.0 0.0 0.0 0.0 0.0 0.0 0.0 0.0 0.0 0.0 0.0 0.0
0.0 0.0 0.0 0.0 0.0 0.0 0.0 0.0 0.0 0.0 0.0 0.0 0.0 0.1 0.9 0.9 0.6 0.4 0.0 0.0 0.0 0.0 0.0 0.0 0.0 0.0 0.0 0.0
0.0 0.0 0.0 0.0 0.0 0.0 0.0 0.0 0.0 0.0 0.0 0.0 0.0 0.0 0.3 0.9 1.0 1.0 0.5 0.1 0.0 0.0 0.0 0.0 0.0 0.0 0.0 0.0
0.0 0.0 0.0 0.0 0.0 0.0 0.0 0.0 0.0 0.0 0.0 0.0 0.0 0.0 0.0 0.2 0.7 1.0 1.0 0.6 0.1 0.0 0.0 0.0 0.0 0.0 0.0 0.0
0.0 0.0 0.0 0.0 0.0 0.0 0.0 0.0 0.0 0.0 0.0 0.0 0.0 0.0 0.0 0.0 0.1 0.4 1.0 1.0 0.7 0.0 0.0 0.0 0.0 0.0 0.0 0.0
0.0 0.0 0.0 0.0 0.0 0.0 0.0 0.0 0.0 0.0 0.0 0.0 0.0 0.0 0.0 0.0 0.0 0.0 1.0 1.0 1.0 0.3 0.0 0.0 0.0 0.0 0.0 0.0
0.0 0.0 0.0 0.0 0.0 0.0 0.0 0.0 0.0 0.0 0.0 0.0 0.0 0.0 0.0 0.0 0.2 0.5 0.7 1.0 1.0 0.8 0.0 0.0 0.0 0.0 0.0 0.0
0.0 0.0 0.0 0.0 0.0 0.0 0.0 0.0 0.0 0.0 0.0 0.0 0.0 0.2 0.6 0.9 1.0 1.0 1.0 1.0 0.7 0.0 0.0 0.0 0.0 0.0 0.0 0.0
0.0 0.0 0.0 0.0 0.0 0.0 0.0 0.0 0.0 0.1 0.4 0.9 1.0 1.0 1.0 1.0 0.8 0.3 0.0 0.0 0.0 0.0 0.0 0.0 0.0 0.0 0.0 0.0
0.0 0.0 0.0 0.0 0.0 0.0 0.0 0.1 0.3 0.8 1.0 1.0 1.0 1.0 0.8 0.3 0.0 0.0 0.0 0.0 0.0 0.0 0.0 0.0 0.0 0.0 0.0 0.0
0.0 0.0 0.0 0.0 0.1 0.7 0.9 1.0 1.0 1.0 1.0 0.8 0.3 0.0 0.0 0.0 0.0 0.0 0.0 0.0 0.0 0.0 0.0 0.0 0.0 0.0 0.0 0.0
0.0 0.0 0.0 0.2 0.7 0.9 1.0 1.0 1.0 1.0 0.5 0.0 0.0 0.0 0.0 0.0 0.0 0.0 0.0 0.0 0.0 0.0 0.0 0.0 0.0 0.0 0.0 0.0
0.0 0.0 0.0 0.5 1.0 1.0 1.0 0.8 0.5 0.5 0.1 0.0 0.0 0.0 0.0 0.0 0.0 0.0 0.0 0.0 0.0 0.0 0.0 0.0 0.0 0.0 0.0 0.0
0.0 0.0 0.0 0.0 0.0 0.0 0.0 0.0 0.0 0.0 0.0 0.0 0.0 0.0 0.0 0.0 0.0 0.0 0.0 0.0 0.0 0.0 0.0 0.0 0.0 0.0 0.0 0.0
0.0 0.0 0.0 0.0 0.0 0.0 0.0 0.0 0.0 0.0 0.0 0.0 0.0 0.0 0.0 0.0 0.0 0.0 0.0 0.0 0.0 0.0 0.0 0.0 0.0 0.0 0.0 0.0
0.0 0.0 0.0 0.0 0.0 0.0 0.0 0.0 0.0 0.0 0.0 0.0 0.0 0.0 0.0 0.0 0.0 0.0 0.0 0.0 0.0 0.0 0.0 0.0 0.0 0.0 0.0 0.0
```

Figure 6.11 – Inspecting a single image matrix

Do you notice how easy it is to spot a 5 in the image? You can execute `train_labels[0]` to verify.

4. You can continue with laying out the neural network architecture next. As mentioned earlier, the input images are 28x28 pixels in size. Artificial neural networks can't process a matrix directly, so you'll have to convert this matrix to a vector. This process is known as **flattening**. As a result, you'll end up with a single vector of size (1, (28x28)), or (1, 784).

    This input data can be passed to our first and only hidden layer, with 128 neurons. You know how to code out neurons and layers manually, but that's inconvenient in practice. Instead, you can use `layers.Dense()` to construct a layer.

This hidden layer will also need an activation function, so you can use ReLU.

Finally, you can add the final (output) layer, which needs to have as many neurons as there are distinct classes – 10 in this case.

Here's the entire code snippet for the network architecture:

```
model = models.Sequential([
    layers.Flatten(input_shape=(28, 28)),
    layers.Dense(128, activation='relu'),
    layers.Dense(10)
])
```

The `models.Sequential` function allows you to stack layers one after the other, and, well, to make a network out of the individual layers.

You can view the architecture of your model by calling the `summary()` method on it:

```
model.summary()
```

The results are shown in the following screenshot:

```
Model: "sequential"
```

Layer (type)	Output Shape	Param #
flatten (Flatten)	(None, 784)	0
dense (Dense)	(None, 128)	100480
dense_1 (Dense)	(None, 10)	1290

```
Total params: 101,770
Trainable params: 101,770
Non-trainable params: 0
```

Figure 6.12 – Neural network architecture

5. There's still one thing you need to do before model training, and that is to compile the model. During the compilation, you'll have to specify values for the optimizer, loss, and the optimization metrics.

These haven't been covered in this chapter, but a brief explanation of each follows:

- *Optimizers* – algorithms used to change the attributes of the neural networks to reduce the loss. These attributes include weights, learning rates, and so on.

- *Loss* – a method used to calculate gradients, which are then used to update the weights in the neural network.

- *Metrics* – the metric(s) you're optimizing for (for example, accuracy).

Going deeper into any of these topics is beyond the scope of this book. There are plenty of resources for discovering the theory behind deep learning. This chapter only aims to cover the essential basics.

You can compile your neural network by executing the following code:

```
model.compile(
    optimizer='adam',
    loss=tf.keras.losses.
SparseCategoricalCrossentropy(from_logits=True),
    metrics=['accuracy']
)
```

6. Now you're ready to train the model. The training subset will be used to train the network, and the testing subset will be used for the evaluation. The network will be trained for 10 epochs (10 complete passes through the entire training data).

    You can use the following code snippet to train the model:

```
history = model.fit(
    train_images,
    train_labels,
    epochs=10,
    validation_data=(test_images, test_labels)
)
```

Executing the preceding code will start the training process. How long it will take depends on the hardware you have and whether you're using a GPU or CPU. You should see something similar to the following screenshot:

```
Epoch 1/10
1875/1875 [==============================] - 1s 477us/step - loss: 0.4476 - accuracy: 0.8761 - val_loss: 0.1400 - val_accuracy: 0.9574
Epoch 2/10
1875/1875 [==============================] - 1s 421us/step - loss: 0.1227 - accuracy: 0.9647 - val_loss: 0.0985 - val_accuracy: 0.9702
Epoch 3/10
1875/1875 [==============================] - 1s 414us/step - loss: 0.0798 - accuracy: 0.9761 - val_loss: 0.0821 - val_accuracy: 0.9757
Epoch 4/10
1875/1875 [==============================] - 1s 414us/step - loss: 0.0598 - accuracy: 0.9818 - val_loss: 0.0739 - val_accuracy: 0.9784
Epoch 5/10
1875/1875 [==============================] - 1s 460us/step - loss: 0.0445 - accuracy: 0.9866 - val_loss: 0.0779 - val_accuracy: 0.9775
Epoch 6/10
1875/1875 [==============================] - 1s 431us/step - loss: 0.0351 - accuracy: 0.9893 - val_loss: 0.0753 - val_accuracy: 0.9776
Epoch 7/10
1875/1875 [==============================] - 1s 420us/step - loss: 0.0255 - accuracy: 0.9923 - val_loss: 0.0777 - val_accuracy: 0.9759
Epoch 8/10
1875/1875 [==============================] - 1s 415us/step - loss: 0.0215 - accuracy: 0.9936 - val_loss: 0.0798 - val_accuracy: 0.9780
Epoch 9/10
1875/1875 [==============================] - 1s 413us/step - loss: 0.0170 - accuracy: 0.9954 - val_loss: 0.0761 - val_accuracy: 0.9787
Epoch 10/10
1875/1875 [==============================] - 1s 412us/step - loss: 0.0146 - accuracy: 0.9957 - val_loss: 0.0851 - val_accuracy: 0.9769
```

Figure 6.13 – MNIST model training

After 10 epochs, the accuracy on the validation set was 97.7% – excellent if we consider that regular neural networks don't work too well with images.

7. To test your model on a new instance, you can use the `predict()` method. It returns an array that tells you how likely it is that the prediction for a given class is correct. There will be 10 items in this array, as there were 10 classes.

    You can then call `np.argmax()` to get the item with the highest value:

```python
import numpy as np

prediction = model.predict(test_images[0].reshape(-1,
784))
print(f'True digit = {test_labels[0]}')
print(f'Predicted digit = {np.argmax(prediction)}')
```

The results are shown in the following screenshot:

```
True digit = 7
Predicted digit = 7
```

Figure 6.14 – Testing the MNIST model

As you can see, the prediction is correct.

And that's how easy it is to train neural networks with libraries such as TensorFlow. Keep in mind that this way of handling image classification isn't recommended in the real world, as we've flattened a 28x28 image and immediately lost all two-dimensional information. CNNs would be a better approach for image classification, as they can extract useful features from two-dimensional data. Our artifical neural network worked well here because MNIST is a simple and clean dataset – not something you'll get a whole lot of in your job.

In the next section, you'll learn the differences in approaching classification and regression tasks with neural networks.

Neural networks in regression versus classification

If you've done any machine learning with scikit-learn, you know there are dedicated classes and models for regression and classification datasets. For example, if you would like to apply a decision tree algorithm to a classification dataset, you would use the `DecisionTreeClassifier` class. Likewise, you would use the `DecisionTreeRegressor` class for regression tasks.

But what do you do with neural networks? There are no dedicated classes or layers for classification and regression tasks.

Instead, you can accommodate by tweaking the number of neurons in the output layer. Put simply, if you're dealing with regression tasks, there has to be a single neuron in the output layer. If you're dealing with classification tasks, there will be as many neurons in the output layer as there are distinct classes in your target variable.

For example, you saw how the neural network in the previous section had 10 neurons in the output layer. The reason is that there are 10 distinct digits, from zero to nine. If you were instead predicting the price of something (regression), there would be only a single neuron in the output layer.

The task of the neural network is to learn the adequate parameter values (weights and biases) to produce the best output value, irrespective of the type of problem you're solving.

Summary

This chapter might be hard to process if this was your first encounter with deep learning and neural networks. Going over the materials a couple of times could help, but it won't be enough to understand the topic fully. Entire books have been written on deep learning, and even on small subsets of deep learning. Hence, covering everything in a single chapter isn't possible.

Still, you should have the basic theory behind the concepts of neurons, layers, and activation functions, and you can always learn more on your own. The following chapter, *Chapter 7, Neural Network Classifier with TPOT*, will show you how to connect neural networks and pipeline optimization, so you can build state-of-the-art models in a completely automated fashion.

As always, please feel free to explore the theory and practice of deep learning and neural networks on your own. It is definitely a field of study worth exploring further.

Q&A

1. How would you define the term "deep learning"?

2. What is the difference between traditional machine learning algorithms and algorithms used in deep learning?

3. List and briefly describe five types of neural networks.

4. Can you figure out how to calculate the number of trainable parameters in a network given the number of neurons per layer? For example, a neural network with the architecture [10, 8, 8, 2] has in total 178 trainable parameters (160 weights and 18 biases).

5. Name four different activation functions and briefly explain them.

6. In your own words, describe *loss* in neural networks.

7. Explain why modeling imagine classification models with regular artificial neural networks isn't a good idea.

7
Neural Network Classifier with TPOT

In this chapter, you'll learn how to build your deep learning classifier in an automated fashion – by using the TPOT library. It's assumed that you know the basics of artificial neural networks, so terms such as *neurons*, *layers*, *activation functions*, and *learning rates* should sound familiar. If you don't know how to explain these terms simply, please revisit *Chapter 6, Getting Started with Deep Learning: Crash Course in Neural Networks*.

Throughout this chapter, you'll learn how easy it is to build a simple classifier based on neural networks and how you can tweak the neural network so that it better suits your needs and the training data.

This chapter will cover the following topics:

- Exploring the dataset
- Exploring options for training neural network classifiers
- Training a neural network classifier

Technical requirements

You don't need any prior hands-on experience with deep learning and neural networks. A knowledge of some basics concepts and terminology is a must, however. If you're entirely new to the subject, please revisit *Chapter 6, Getting Started with Deep Learning: Crash Course in Neural Networks.*

You can download the source code and dataset for this chapter here: `https://github.com/PacktPublishing/Machine-Learning-Automation-with-TPOT/tree/main/Chapter07`.

Exploring the dataset

There is no reason to go wild with the dataset. Just because we can train neural network models with TPOT doesn't mean we should spend 50+ pages exploring and transforming needlessly complex datasets.

For that reason, you'll use a scikit-learn built-in dataset throughout the chapter – the Breast cancer dataset. This dataset doesn't have to be downloaded from the web as it comes built-in with scikit-learn. Let's start by loading and exploring it:

1. To begin, you'll need to load in a couple of libraries. We're importing NumPy, pandas, Matplotlib, and Seaborn for easy data analysis and visualization. Also, we're importing the `load_breast_cancer` function from the `sklearn.datasets` module. That's the function that will load in the dataset. Finally, the `rcParams` module is imported from Matplotlib to make default styling a bit easier on the eyes:

    ```
    import numpy as np
    import pandas as pd
    import matplotlib.pyplot as plt
    import seaborn as sns
    from sklearn.datasets import load_breast_cancer

    from matplotlib import rcParams
    rcParams['figure.figsize'] = (14, 7)
    rcParams['axes.spines.top'] = False
    rcParams['axes.spines.right'] = False
    ```

2. You can now use the `load_breast_cancer` function to load in the dataset. The function returns a dictionary, so we can use the `keys()` method to print the keys:

    ```
    data = load_breast_cancer()
    data.keys()
    ```

The results are shown in the following diagram:

```
dict_keys(['data', 'target', 'frame', 'target_names', 'DESCR', 'feature_names', 'filename'])
```

Figure 7.1 – Dictionary keys of the Breast cancer dataset

3. You can now use this dictionary to extract attributes of interest. What is essential for now are the `data` and `target` keys. You can store their values to separate variables and then construct a data frame object from them. Working with raw values is possible, but a pandas data frame data structure will allow easier data manipulation, transformation, and exploration.

    Here's how you can transform these to a pandas data frame:

    ```
    features = data.data
    target = data.target

    df =\
    pd.DataFrame(data=features,columns=data.feature_names)
    df['target'] = target
    df.sample(8)
    ```

    The results are shown in the following table:

	mean radius	mean texture	mean perimeter	mean area	mean smoothness	mean compactness	mean concavity	mean concave points	mean symmetry	mean fractal dimension	...
37	13.03	18.42	82.61	523.8	0.08983	0.03766	0.025620	0.029230	0.1467	0.05863	...
104	10.49	19.29	67.41	336.1	0.09989	0.08578	0.029950	0.012010	0.2217	0.06481	...
129	19.79	25.12	130.40	1192.0	0.10150	0.15890	0.254500	0.114900	0.2202	0.06113	...
481	13.90	19.24	88.73	602.9	0.07991	0.05326	0.029950	0.020700	0.1579	0.05594	...
105	13.11	15.56	87.21	530.2	0.13980	0.17650	0.207100	0.096010	0.1925	0.07692	...
155	12.25	17.94	78.27	460.3	0.08654	0.06679	0.038850	0.023310	0.1970	0.06228	...
303	10.49	18.61	66.86	334.3	0.10680	0.06678	0.022970	0.017800	0.1482	0.06600	...
522	11.26	19.83	71.30	388.1	0.08511	0.04413	0.005067	0.005664	0.1637	0.06343	...

8 rows × 31 columns

Figure 7.2 – Sample of eights rows from the Breast cancer dataset

4. The first thing you always want to do, analysis-wise, is to check for missing data. Pandas has an `isnull()` method built in, which returns Booleans for every value in the dataset. You can then call the `sum()` method on top of these results to get the count of missing values per column:

    ```
    df.isnull().sum()
    ```

The results are shown in the following diagram:

```
mean radius                 0
mean texture                0
mean perimeter              0
mean area                   0
mean smoothness             0
mean compactness            0
mean concavity              0
mean concave points         0
mean symmetry               0
mean fractal dimension      0
radius error                0
texture error               0
perimeter error             0
area error                  0
smoothness error            0
compactness error           0
concavity error             0
concave points error        0
symmetry error              0
fractal dimension error     0
worst radius                0
worst texture               0
worst perimeter             0
worst area                  0
worst smoothness            0
worst compactness           0
worst concavity             0
worst concave points        0
worst symmetry              0
worst fractal dimension     0
target                      0
dtype: int64
```

Figure 7.3 – Missing value count per column

As you can see, there are no missing values.

5. The next thing to do in the exploratory phase is to get familiar with your dataset. Data visualization techniques can provide an excellent way of doing so.

For example, you can declare a function called `make_count_chart()` that takes any categorical attribute and visualizes its distribution. Here's what the code for this function could look like:

```
def make_count_chart(column, title, ylabel, xlabel, y_
offset=0.12, x_offset=700):
    ax = df[column].value_counts().plot(kind='bar',
fontsize=13, color='#4f4f4f')
    ax.set_title(title, size=20, pad=30)
    ax.set_ylabel(ylabel, fontsize=14)
    ax.set_xlabel(xlabel, fontsize=14)

    for i in ax.patches:
        ax.text(i.get_x() + x_offset, i.get_height()\
```

```
        + y_offset, f'{str(round(i.get_height(), 2))}',\
    fontsize=15)
        return ax
```

You can now use the following code snippet to visualize the target variable to find out how many instances were benign and how many were malignant:

```
make_count_chart(
        column='target',
        title=\
    'Number of malignant (1) vs benign (0) cases',
        ylabel='Malignant? (0 = No, 1 = Yes)',
        xlabel='Count',
        y_offset=10,
        x_offset=0.22
    )
```

The results are shown in the following diagram:

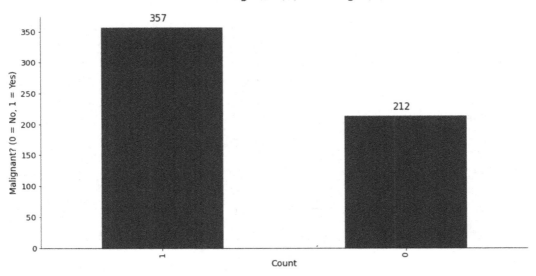

Figure 7.4 – Number of malignant and benign cases

As you can see, there's a decent amount more of malignant cases, so the classes aren't perfectly balanced. Class imbalance can lead to highly accurate but unusable models. Just imagine you are classifying a rare event. In every 10,000 transactions, only one is classified as an anomaly. Clearly, the machine learning model doesn't have much chance to learn what makes an anomaly so different from the rest.

Furthermore, always predicting that the transaction is normal leads to a 99.99% accurate model. State-of-the-art accuracy, for certain, but the model is unusable.

There are numerous techniques for dealing with imbalanced datasets, but they are beyond the scope for this book.

6. Next stop – correlation analysis. The aim of this step is to take a glimpse at which feature(s) have the biggest effect on the target variables. In other words, we want to establish how correlated a change in direction in a feature is with the target class. Visualizing an entire correlation matrix on a dataset of 30+ columns isn't the best idea because it would require a figure too large to fit comfortably on a single page. Instead, we can calculate the correlation of the feature with the target variable.

    Here's how you can do this for the mean area feature – by calling the corrcoeff() method from NumPy:

```
np.corrcoef(df['mean area'], df['target'])[1][0]
```

    The results are shown in the following figure:

$$-0.7089838365853899$$

Figure 7.5 – Correlation coefficient between a single feature and the target variable

You can now easily apply the same logic to the entire dataset. The following code snippet keeps track of the correlation between every feature and the target variable, converts the results to a pandas data frame, and sorts the values by the correlation coefficient in descending order:

```
corr_with_target = []

for col in df.columns[:-1]:
    corr = np.corrcoef(df[col], df['target'])[1][0]
    corr_with_target.append({'Column': col,
'Correlation': corr})

corr_df = pd.DataFrame(corr_with_target)
corr_df = \
corr_df.sort_values(by='Correlation', ascending=False)
```

Please note the [:-1] at the beginning of the loop. Since the target variable is the last column, we can use the aforementioned slicing technique to exclude the target variable from the correlation calculation. The correlation coefficient between the target variable and the non-target variable would be 1, which is not particularly useful to us.

You can now use the following code to make a horizontal bar chart of the correlations with the target variable:

```
plt.figure(figsize=(10, 14))
plt.barh(corr_df['Column'], corr_df['Correlation'],
color='#4f4f4f')
plt.title('Feature correlation with the target variable',
fontsize=20)
plt.xlabel('Feature', fontsize=14)
plt.ylabel('Correlation', fontsize=14)
plt.show()
```

The results are shown in the following diagram:

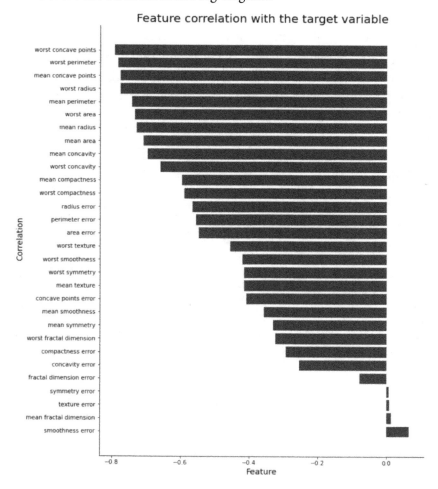

Figure 7.6 – Feature correlation with the target variable

As you can see, most of the features have a high negative correlation with the target variable. Negative correlation means that one variable increases as the other one decreases. In our case, a decrease in the number of features leads to an increase in the target variable.

7. You could also visualize the distribution of every numeric column with respect to the target variable value. To be more precise, this means two separate histograms are drawn on a single chart, and each histogram shows the distribution only for the respective target value's subset.

For example, this means that one histogram will show the distribution of malignant and the other of benign instances, for each variable.

The code snippet you're about to see declares a `draw_histogram()` function that goes over every column in a dataset, makes a histogram with respect to the distinct classes in the target variable, and appends this histogram to a figure.

Once all of the histograms are appended, the figure is displayed to the user. The user also has to specify how many rows and columns they want, which gives a bit of extra freedom when designing visualizations.

Here is the code snippet for drawing this histogram grid:

```python
def draw_histogram(data, columns, n_rows, n_cols):
    fig = plt.figure(figsize=(12, 18))
    for i, var_name in enumerate(columns):
        ax = fig.add_subplot(n_rows, n_cols, i + 1)
        sns.histplot(data=data, x=var_name, hue='target')
        ax.set_title(f'Distribution of {var_name}')
    fig.tight_layout()
    plt.show()

draw_histogram(df, df.columns[:-1], 9, 4)
```

This will be a pretty large data visualization, containing 9 rows and 4 columns. The last row will have only 2 histograms, as there are 30 continuous variables in total.

The results are shown in the following diagram:

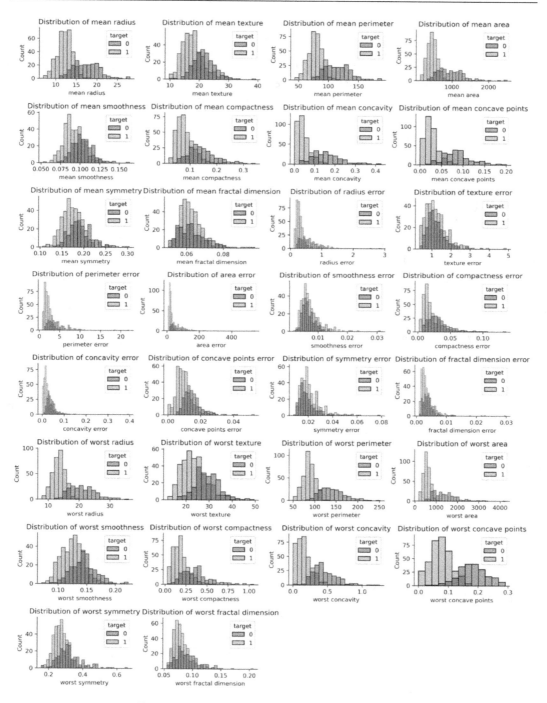

Figure 7.7 – Histogram for every continuous variable

As you can see, there is a distinct separation most of the time, so our model shouldn't have too much trouble making decent separations between the classes.

And that's all we'll do with regard to the exploratory data analysis. You can, and are encouraged to, do more, especially with custom and more complex datasets. The next section will introduce you to the options you have for training automated neural network classifiers.

Exploring options for training neural network classifiers

You have a lot of options when training neural network models with TPOT. The whole neural network story is still new and experimental with TPOT, requiring a bit more manual work than regular scikit-learn estimators.

By default, TPOT won't use the neural network models unless you explicitly specify that it has to. This specification is done by selecting an adequate configuration dictionary that includes one or more neural network estimators (you can also write these manually).

The more convenient option is to import configuration dictionaries from the `tpot/config/classifier_nn.py` file. This file contains two PyTorch classifier configurations, as visible in the following diagram:

```
'tpot.builtins.PytorchLRClassifier': {
    'learning_rate': [1e-3, 1e-2, 1e-1, 0.5, 1.],
    'batch_size': [4, 8, 16, 32],
    'num_epochs': [5, 10, 15],
    'weight_decay': [0, 1e-4, 1e-3, 1e-2]
},

'tpot.builtins.PytorchMLPClassifier': {
    'learning_rate': [1e-3, 1e-2, 1e-1, 0.5, 1.],
    'batch_size': [4, 8, 16, 32],
    'num_epochs': [5, 10, 15],
    'weight_decay': [0, 1e-4, 1e-3, 1e-2]
},
```

Figure 7.8 – TPOT PyTorch classifier configurations

From the preceding diagram, you can see that TPOT can currently handle two different types of classifiers based on deep learning libraries:

- Logistic regression: shown in `tpot.builtins.PytorchLRClassifier`
- Multi-layer perceptron: shown in `tpot.builtins.PytorchMLPClassifier`

You can either import this file or write the configurations manually. In addition, you can also specify your own configuration dictionaries, which somehow modify the existing ones. For example, you can use this code to use a PyTorch-based logistic regression estimator:

```
tpot_config = {
    'tpot.nn.PytorchLRClassifier': {
        'learning_rate': [1e-3, 1e-2, 1e-1, 0.5, 1.]
    }
}
```

Custom configurations will be discussed later in the chapter when we start to implement neural network classifiers.

You should keep in mind that training neural network classifiers with TPOT is an expensive task and will typically take much more time to train than scikit-learn estimators. As a rule of thumb, you should expect the training time to be several orders of magnitude slower with neural networks. This is because neural network architectures can have millions of trainable and adjustable parameters, and finding the correct value for all of them takes time.

Having that in mind, you should always consider simpler options first, as TPOT is highly likely to give you an excellent-performing pipeline on the default scikit-learn estimators.

The next section will continue with the training of neural network classifiers right where the previous section stopped and will show you how different training configurations can be used to train your models.

Training a neural network classifier

Up to this point, we've loaded in the dataset and undertaken a basic exploratory data analysis. This section of the chapter will focus on training models through different configurations:

1. Before we can move on to model training, we need to split our dataset into training and testing subsets. Doing so will allow us to have a sample of the data never seen by the model, and which can later be used for evaluation.

The following code snippet will split the data in a 75:25 ratio:

```
from sklearn.model_selection import train_test_split

X = df.drop('target', axis=1)
y = df['target']

X_train, X_test, y_train, y_test =train_test_split(\
X, y, test_size=0.25, random_state=42)
```

We can begin with training next.

2. As always, let's start simply by training a baseline model. This will serve as a minimum viable performance that the neural network classifier has to outperform.

The simplest binary classification algorithm is logistic regression. The following code snippet imports it from scikit-learn, alongside some evaluation metrics, such as a confusion matrix and an accuracy score. Furthermore, the snippet instantiates the model, trains it, makes a prediction on the holdout set, and prints the confusion matrix and the accuracy score.

The code snippet is as follows:

```
from sklearn.linear_model import LogisticRegression
from sklearn.metrics import confusion_matrix, accuracy_
score

lr_model = LogisticRegression()
lr_model.fit(X_train, y_train)
lr_preds = lr_model.predict(X_test)

print(confusion_matrix(y_test, lr_preds))
print()
print(accuracy_score(y_test, lr_preds))
```

The results are shown in the following diagram:

```
[[50  4]
 [ 1 88]]

0.965034965034965
```

Figure 7.9 – Confusion matrix and accuracy of the baseline model

We now know that the baseline model is 96.5% accurate, making 4 false positives and 1 false negative. Next, we'll train an automated neural network classifier with TPOT and see how the results compare.

3. As mentioned before, training a neural network classifier with TPOT is a heavy task. For that reason, you might be better off switching to a free GPU Cloud environment, such as *Google Colab*.

    This will ensure faster training time, but also you won't melt your PC. Once there, you can use the following snippet to train the PyTorch-based logistic regression model:

```
from tpot import TPOTClassifier

classifier_lr = TPOTClassifier(
    config_dict='TPOT NN',
    template='PytorchLRClassifier',
    generations=2,
    random_state=42,
    verbosity=3
)


classifier_lr.fit(X_train, y_train)
```

This will train the model for two generations. You'll see various outputs during training, such as the following:

```
34 operators have been imported by TPOT.
```

Optimization Progress: 17%   52/300 [04:03<23:57, 5.80s/pipeline]

Figure 7.10 – TPOT neural network training process

The training process will take a long time. Please make sure that your PC doesn't go into sleep mode, and also make sure to click somewhere on a blank space in Google Colab every couple of minutes. Failing to do so will cause runtime disconnection, and you'll have to start training from the beginning.

This happens because Colab is a free environment, and if you aren't active, it's assumed you're not using it, so the resources you're occupying are allocated to someone else.

Once the training process completes, you'll see the following output on your screen:

```
34 operators have been imported by TPOT.
Pipeline encountered that has previously been evaluated during the optimization process. Using the score from the

Generation 1 - Current Pareto front scores:

-1      0.9153214774281807      PytorchLRClassifier(input_matrix, PytorchLRClassifier__batch_size=32, PytorchLRCla
Pipeline encountered that has previously been evaluated during the optimization process. Using the score from the
Pipeline encountered that has previously been evaluated during the optimization process. Using the score from the
Pipeline encountered that has previously been evaluated during the optimization process. Using the score from the
Pipeline encountered that has previously been evaluated during the optimization process. Using the score from the
Pipeline encountered that has previously been evaluated during the optimization process. Using the score from the

Generation 2 - Current Pareto front scores:

-1      0.9153214774281807      PytorchLRClassifier(input_matrix, PytorchLRClassifier__batch_size=32, PytorchLRCla
TPOTClassifier(config_dict='TPOT NN', crossover_rate=0.1, cv=5,
               disable_update_check=False, early_stop=None, generations=2,
               log_file=None, max_eval_time_mins=5, max_time_mins=None,
               memory=None, mutation_rate=0.9, n_jobs=1, offspring_size=None,
               periodic_checkpoint_folder=None, population_size=100,
               random_state=42, scoring=None, subsample=1.0,
               template='PytorchLRClassifier', use_dask=False, verbosity=3,
               warm_start=False)
```

Figure 7.11 – TPOT PyTorch logistic regression classifier

The pipeline was only optimized for two generations, so please don't expect any groundbreaking results. Set it to a realistic number if you want a better performing model.

Once the training is done, you can use the following line of code to examine the best-fitted pipeline:

```
classifier_lr.fitted_pipeline_
```

The results are shown in the following diagram:

```
Pipeline(memory=None,
         steps=[('pytorchlrclassifier',
                 PytorchLRClassifier(batch_size=32, learning_rate=0.5,
                                     num_epochs=15, verbose=False,
                                     weight_decay=0.0001))],
         verbose=False)
```

Figure 7.12 – TPOT PyTorch logistic regression best pipeline

You can now export this pipeline if you want to, but we won't do this in this section.

Before proceeding to the next model, let's quickly examine how the PyTorch logistic regression model performed. Just like with the baseline model, we'll print the confusion matrix and the accuracy score.

You can use the following code snippet to do precisely that:

```
from sklearn.metrics import confusion_matrix, \
  accuracy_score

tpot_lr_preds = classifier_lr.predict(X_test)

print(confusion_matrix(y_test, tpot_lr_preds))
print()
print(accuracy_score(y_test, tpot_lr_preds))
```

The results are shown in the following figure:

```
[[40 14]
 [ 0 89]]
```

```
0.9020979020979021
```

Figure 7.13 – Confusion matrix and accuracy score of a PyTorch logistic regression model

As you can see, two generations weren't enough to produce a better-than-baseline model. Let's see whether using a multi-layer perceptron model could help.

4. We're still in the Google Colab environment, as training on your own PC is significantly slower (depending on your configuration). The idea now is to use the multi-layer perceptron model instead of logistic regression and see how the change in the model could affect performance.

    To begin, you'll have to make a change to the `template` parameter of `TPOTClassifier`, as shown here:

    ```
    classifier_mlp = TPOTClassifier(
        config_dict='TPOT NN',
        template='PytorchMLPClassifier',
        generations=2,
        random_state=42,
        verbosity=3
    )
    ```

    As you can see, we're now using `PytorchMLPClassifier` instead of `PytorchLRClassifier`. To begin the optimization process, simply call the `fit()` method with the training data:

    ```
    classifier_mlp.fit(X_train, y_train)
    ```

As with the logistic regression algorithm, you'll also see the progress bar during the optimization process:

34 operators have been imported by TPOT.

Optimization Progress: 54%   162/300 [21:25<32:05, 13.95s/pipeline]

Figure 7.14 – TPOT multi-layer perceptron training process

Once the training process is complete, you'll be presented with the following results:

```
34 operators have been imported by TPOT.
Pipeline encountered that has previously been evaluated during the optimization process. Using the score

Generation 1 - Current Pareto front scores:

-1      0.8988782489740081      PytorchMLPClassifier(input_matrix, PytorchMLPClassifier__batch_size=16,
Pipeline encountered that has previously been evaluated during the optimization process. Using the score
Pipeline encountered that has previously been evaluated during the optimization process. Using the score
Pipeline encountered that has previously been evaluated during the optimization process. Using the score

Generation 2 - Current Pareto front scores:

-1      0.8988782489740081      PytorchMLPClassifier(input_matrix, PytorchMLPClassifier__batch_size=16,
TPOTClassifier(config_dict='TPOT NN', crossover_rate=0.1, cv=5,
               disable_update_check=False, early_stop=None, generations=2,
               log_file=None, max_eval_time_mins=5, max_time_mins=None,
               memory=None, mutation_rate=0.9, n_jobs=1, offspring_size=None,
               periodic_checkpoint_folder=None, population_size=100,
               random_state=42, scoring=None, subsample=1.0,
               template='PytorchMLPClassifier', use_dask=False, verbosity=3,
               warm_start=False)
```

Figure 7.15 – TPOT PyTorch multi-layer perceptron classifier

Once again, the pipeline was only optimized for two generations, so please don't expect any groundbreaking results. Set it to a realistic number if you want a better performing model.

Once the training is complete, you can use the following line of code to examine the best-fitted pipeline:

```
classifier_mlp.fitted_pipeline_
```

The results are shown in the following diagram:

```
Pipeline(memory=None,
         steps=[('pytorchmlpclassifier',
                 PytorchMLPClassifier(batch_size=16, learning_rate=0.001,
                                      num_epochs=15, verbose=False,
                                      weight_decay=0.001))],
         verbose=False)
```

Figure 7.16 – TPOT PyTorch multi-layer perceptron best pipeline

Let's quickly examine how the logistic regression model with PyTorch performed. Just like with the previous model, we'll print the confusion matrix and the accuracy score.

You can use the following code snippet to do precisely that:

```
from sklearn.metrics import confusion_matrix,\
  accuracy_score

tpot_mlp_preds = classifier_mlp.predict(X_test)

print(confusion_matrix(y_test, tpot_mlp_preds))
print()
print(accuracy_score(y_test, tpot_mlp_preds))
```

The results are shown in the following figure:

```
[[48  6]
 [ 3 86]]

0.9370629370629371
```

Figure 7.17 – Confusion matrix and accuracy score of a PyTorch multi-layer perceptron model

As you can see, two generations still weren't enough to produce a better-than-baseline model, but the MLP model outperformed the logistic regression one. Let's now see whether using a custom training configuration could push the accuracy even higher.

5. Finally, let's see how you can specify possible hyperparameter values for either logistic regression or multi-layer perceptron models. All you have to do is specify a custom configuration dictionary, which holds the hyperparameters you want to test for (such as learning rate, batch size, and number of epochs), and assign values to those hyperparameters in the form of a list.

    Here's an example:

```
custom_config = {
    'tpot.builtins.PytorchMLPClassifier': {
        'learning_rate': [1e-1, 0.5, 1.],
        'batch_size': [16, 32],
        'num_epochs': [10, 15],
    }
}
```

You can now use this custom_config dictionary when training models. Here is an example training snippet based on a multi-layer perceptron model:

```
classifier_custom = TPOTClassifier(
    config_dict=custom_config,
    template='PytorchMLPClassifier',
```

```
        generations=2,
        random_state=42,
        verbosity=3
)

classifier_custom.fit(X_train, y_train)
```

As you can see, only the `config_dict` parameter has changed. Once the training process has started, you'll see a progress bar similar to this one in the notebook:

1 operators have been imported by TPOT.

Optimization Progress: 32%   **96/300 [01:18<30:37, 9.01s/pipeline]**

Figure 7.18 – TPOT custom tuning with neural networks

Once the training process is complete, you should see something along the following lines in the notebook:

```
1 operators have been imported by TPOT.
Pipeline encountered that has previously been evaluated during the optimization process. Using the score from the previous evaluation.
Pipeline encountered that has previously been evaluated during the optimization process. Using the score from the previous evaluation.
Pipeline encountered that has previously been evaluated during the optimization process. Using the score from the previous evaluation.
Pipeline encountered that has previously been evaluated during the optimization process. Using the score from the previous evaluation.
Pipeline encountered that has previously been evaluated during the optimization process. Using the score from the previous evaluation.
Pipeline encountered that has previously been evaluated during the optimization process. Using the score from the previous evaluation.
Pipeline encountered that has previously been evaluated during the optimization process. Using the score from the previous evaluation.
Pipeline encountered that has previously been evaluated during the optimization process. Using the score from the previous evaluation.
Pipeline encountered that has previously been evaluated during the optimization process. Using the score from the previous evaluation.
Pipeline encountered that has previously been evaluated during the optimization process. Using the score from the previous evaluation.
Pipeline encountered that has previously been evaluated during the optimization process. Using the score from the previous evaluation.
Pipeline encountered that has previously been evaluated during the optimization process. Using the score from the previous evaluation.

Generation 1 - Current Pareto front scores:

-1      0.6291108071135431      PytorchMLPClassifier(input_matrix, PytorchMLPClassifier__batch_size=16, PytorchMLPClassifier__learning
Pipeline encountered that has previously been evaluated during the optimization process. Using the score from the previous evaluation.
Pipeline encountered that has previously been evaluated during the optimization process. Using the score from the previous evaluation.
Pipeline encountered that has previously been evaluated during the optimization process. Using the score from the previous evaluation.
Pipeline encountered that has previously been evaluated during the optimization process. Using the score from the previous evaluation.
Pipeline encountered that has previously been evaluated during the optimization process. Using the score from the previous evaluation.
Pipeline encountered that has previously been evaluated during the optimization process. Using the score from the previous evaluation.
Pipeline encountered that has previously been evaluated during the optimization process. Using the score from the previous evaluation.
Pipeline encountered that has previously been evaluated during the optimization process. Using the score from the previous evaluation.
Pipeline encountered that has previously been evaluated during the optimization process. Using the score from the previous evaluation.
Pipeline encountered that has previously been evaluated during the optimization process. Using the score from the previous evaluation.

Generation 2 - Current Pareto front scores:

-1      0.6291108071135431      PytorchMLPClassifier(input_matrix, PytorchMLPClassifier__batch_size=16, PytorchMLPClassifier__learning
TPOTClassifier(config_dict={'tpot.builtins.PytorchMLPClassifier': {'batch_size': [16,
                                                                                32],
                                                   'learning_rate': [0.1,
                                                                     0.5,
                                                                     1.0],
                                                   'num_epochs': [10,
                                                                 15]}},
              crossover_rate=0.1, cv=5, disable_update_check=False,
              early_stop=None, generations=2, log_file=None,
              max_eval_time_mins=5, max_time_mins=None, memory=None,
              mutation_rate=0.9, n_jobs=1, offspring_size=None,
              periodic_checkpoint_folder=None, population_size=100,
              random_state=42, scoring=None, subsample=1.0,
              template='PytorchMLPClassifier', use_dask=False, verbosity=3,
              warm_start=False)
```

Figure 7.19 – TPOT multi-layer perceptron classifier with custom hyperparameters

And that's all there is to it! Just to verify, you can examine the best-fitted pipeline by executing the following command:

```
classifier_custom.fitted_pipeline_
```

The results are shown in the following figure:

```
Pipeline(memory=None,
        steps=[('pytorchmlpclassifier',
                PytorchMLPClassifier(batch_size=16, learning_rate=0.5,
                                    num_epochs=10, verbose=False,
                                    weight_decay=0))],
        verbose=False)
```

Figure 7.20 – TPOT best-fitted pipeline for a model with custom hyperparameters

As you can see, all of the hyperparameter values are within the specified range, which indicates that the custom model was trained successfully.

This concludes the model training portion of this section and this section in general. What follows is a brief summary of everything we have learned thus far, and a brief introduction to everything that will follow in the upcoming chapters.

Summary

This chapter was quite intensive in terms of hands-on examples and demonstrations. You've hopefully managed to learn how to train automated classification pipelines with TPOT and what you can tweak during the process.

You should now be capable of training any kind of automated machine learning model with TPOT, whether we're talking about regression, classification, standard classifiers, or neural network classifiers. There is good news, as this was the last chapter with TPOT examples.

In the following chapter, *Chapter 8, TPOT Model Deployment*, you'll learn how to wrap the predictive functionality of your models inside a REST API, which will then be tested and deployed both locally and to the cloud. You'll also learn how to communicate with the API once it's deployed.

Finally, in the previous chapter, *Chapter 9, Using the Deployed TPOT Model in Production*, you'll learn how to develop something useful with the deployed APIs. To be more precise, you'll learn how to make predictions in the Notebook environment by making REST calls to the deployed API, and you'll learn how to develop a simple GUI application that makes your model presentable to the end user.

As always, feel free to study TPOT in more depth, but by now, you're well ahead of the majority, and you're ready to make machine learning useful. See you there!

Questions

1. Which two algorithms are available in TPOT with regard to neural networks?

2. Approximately how many times are neural network classifiers slower to train than the default, scikit-learn ones?

3. List and briefly explain the different hyperparameters available when training models with TPOT and neural networks.

4. Can you specify a custom range of hyperparameter values when training custom neural network models with TPOT? If so, how?

5. How can you find the best-fitted pipeline after the model has finished training?

6. What are the advantages of using a GPU runtime such as Google Colab when training neural network models with TPOT?

7. Describe why a single neuron in the multi-layer perceptron model can be thought of as logistic regression.

8
TPOT Model Deployment

In this chapter, you'll learn how to deploy any automated machine learning model, both to localhost and the cloud. You'll learn why the deployment step is necessary if you aim to make machine learning-powered software. It's assumed you know how to train basic regression and classification models with TPOT. Knowledge of the topics of the last couple of chapters (Dask and neural networks) isn't required, as we won't deal with those here.

Throughout the chapter, you'll learn how easy it is to wrap your models in an API and expose their predictive capabilities to other users that aren't necessarily data scientists. You'll also learn which cloud providers are the best to get you started entirely for free.

This chapter will cover the following topics:

- Why do we need model deployment?
- Introducing `Flask` and `Flask-RESTful`
- Best practices for deploying automated models
- Deploying machine learning models to localhost
- Deploying machine learning models to the cloud

Technical requirements

As briefly said earlier, you need to know how to build basic machine learning models with TPOT. Don't worry if you don't feel too comfortable with the library yet, as we'll develop the model from scratch. If you're entirely new to TPOT, please refer to *Chapter 2, Deep Dive into TPOT, Chapter 3, Exploring Regression with TPOT*, and *Chapter 4, Exploring Classification with TPOT*.

This chapter will be quite code-heavy, so you can refer to the official GitHub repository if you get stuck: `https://github.com/PacktPublishing/Machine-Learning-Automation-with-TPOT/tree/main/Chapter08`.

Why do we need model deployment?

If you're already going through the hassle of training and optimizing machine learning models, why don't you take it a step further and deploy it so everyone can use it?

Maybe you want to have the model's predictive capabilities available in a web application. Perhaps you're a mobile app developer who wants to bring machine learning to Android and iOS. The options are endless and different, but all of them share one similarity – the need to be deployed.

Now, machine learning model deployment has nothing to do with machine learning. The aim is to write a simple REST API (preferably in Python, since that's the language used throughout the book) and expose any form of endpoint that calls a `predict()` function to the world. You want parameters sent to your application in JSON format, and then to use them as inputs to your model. Once the prediction is made, you can simply return it to the user.

Yes, that's all there is to machine learning model deployment. Of course, things could get more technical, but keeping things simple will get us 95% of the way, and you can always explore further to squeeze that extra 5%.

When it comes to the technical side of model deployment, Python provides you with a bunch of options. You can use either `Flask` and `Flask-RESTful`, `FastAPI`, `Django`, or `Pyramid`. There are other options, sure, but their "market share" is more or less negligible. You'll use the first option in this chapter, starting from the next section.

The section that follows aims to introduce you to the libraries with a couple of basic hands-on examples. We'll dive into machine learning afterward.

Introducing Flask and Flask-RESTful

`Flask` is a lightweight framework for building web applications. It enables you to start simple and scale when needed. `Flask-RESTful` is an extension for `Flask` that allows you to build REST APIs in no time.

To get started with these two, you'll need to install them. You can execute the following line from the terminal:

```
> pip install flask flask-restful
```

And that's all you need to get started. Let's explore the basics of `Flask` first:

1. Believe it or not, you'll need only seven lines of code to create your first web application with `Flask`. It won't do anything useful, but it's still a step in the right direction.

    To start, you'll need to import the library and create an application instance. You'll then have to make a function that returns what you want to be displayed on the website and decorate the function with an `@app.route(route_url)` decorator. Keep in mind that you should replace `route_url` with the URL string at which the function should display the result. If you pass in a forward slash (/), the results will be displayed on the root page – but more about that in a bit.

    Finally, you'll have to make the Python file runnable with an `if __name__ == '__main__'` check. The application will run on localhost on port `8000`.

    Refer to the following code snippet for your first `Flask` application:

    ```python
    from flask import Flask

    app = Flask(__name__)

    @app.route('/')
    def root():
        return 'Your first Flask app!'

    if __name__ == '__main__':
        app.run(host='0.0.0.0', port=8000)
    ```

To run the application, you'll have to execute the Python file from the terminal. The file is named `flask_basics.py` on my machine, so to run it, please execute the following:

```
> python flask_basics.py
```

If you did everything correctly, you'll see the following message displayed in your terminal window:

```
* Environment: production
  WARNING: This is a development server. Do not use it in a production deployment.
  Use a production WSGI server instead.
* Debug mode: off
* Running on http://0.0.0.0:8000/ (Press CTRL+C to quit)
127.0.0.1 - - [21/Feb/2021 09:42:44] "GET / HTTP/1.1" 200 -
127.0.0.1 - - [21/Feb/2021 09:42:44] "GET /favicon.ico HTTP/1.1" 404 -
```

Figure 8.1 – Running your first Flask application

From the `Running on http://0.0.0.0:8000/` message, you can see where the application is running, ergo which URL you need to visit to see your application. Just so you know, the `0.0.0.0` part can be replaced with `localhost`.

Once there, you'll see the following displayed, indicating that everything worked properly:

Your first Flask app!

Figure 8.2 – Your first Flask application

And that's how easy it is to build your first web application with `Flask`. You'll learn how to make something a bit more complex next.

2. By now, you know how to build the simplest application with `Flask` – but that's not why you're here. We want to make APIs instead of apps, and that's a bit of a different story. The difference between them is quite obvious – APIs don't come with a user interface (except for the documentation page), whereas web apps do. APIs are just a service. As it turns out, you can build APIs with `Flask` out of the box. We'll explore how to do so and explain why it isn't the best option.

To start, create a new Python file. This file will be referenced as `flask_basics2.py`. Inside it, we'll have a single route for two possible API call types. Both have the task of adding two numbers and returning the results, but they do so differently. Let's list the differences:

a) `/adding` (GET) relies on the logic implemented earlier. To be more precise, a GET request is made when the endpoint is called. The only difference is that this time, the parameters are passed in the URL. For example, calling `/adding?num1=3&num2=5` should display 8 onscreen. The parameter values are extracted directly from the URL. You'll see this in action, so everything will be clear immediately.

b) `/adding` (POST) is quite similar to the first endpoint but makes a POST request instead. This is a more secure communication method, as parameter values aren't passed in the URL directly but in the request body instead. This endpoint returns the summation as JSON, so you'll need to wrap the results inside the `flask.jsonify()` function.

Both functions aim to complete an identical task – sum two numbers and return the result. Here's one example of how you might implement this logic:

```python
from flask import Flask, request, jsonify

app = Flask(__name__)

@app.route('/adding')
def add_get():
    num1 = int(request.args.get('num1'))
    num2 = int(request.args.get('num2'))
    return f'<h3>{num1} + {num2} = {num1 + num2}</h3>'


@app.route('/adding', methods=['POST'])
def add_post():
    data = request.get_json()
    num1 = data['num1']
    num2 = data['num2']
    return jsonify({'result': num1 + num2})
```

```
if __name__ == '__main__':
    app.run(host='0.0.0.0', port=8000)
```

As you can see, the `add_get()` function returns a string formatted as HTML. You can return entire HTML documents if you want, but that's not something that interests us now, so we won't look further into it.

To run the application, you'll have to execute the Python file from the terminal. The file is named `flask_basics2.py` on my machine, so to run it, please execute the following:

```
> python flask_basics2.py
```

The API will now run, and you can see both endpoints. Let's take a look at `/adding` for GET first:

Internal Server Error

The server encountered an internal error and was unable to complete your request. Either the server is overloaded or there is an error in the application.

Figure 8.3 – The GET endpoint without parameter values

As you can see, calling the endpoint but failing to provide parameter values will result in an error. There's a way around this, but we're not too interested in GET methods in this chapter, so we'll leave it as is.

We could continue testing the API from the browser, but there's a better option – Postman. This is a free application designed for API testing. You can download it from here: `https://www.postman.com`. The installation is straightforward – simply download the installation file corresponding to your OS, specify the install location, and agree to the license terms. Don't feel obligated to change anything; just leave every setting as is by default.

Once installed, let's open the Postman app and enter the following URL: `http://localhost:8000/adding?num1=3&num2=5`. The **Params** tab will immediately populate, and after that, you're ready to send the request.

You should see the following output almost immediately:

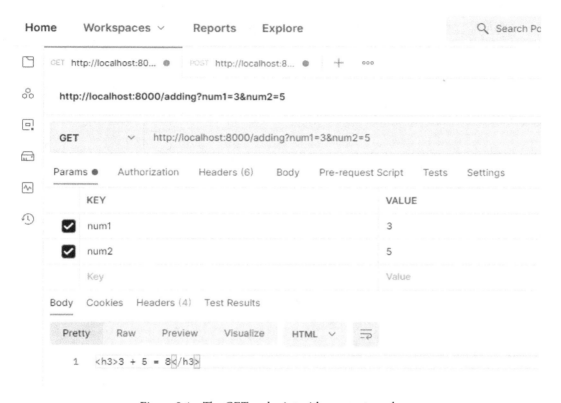

Figure 8.4 – The GET endpoint with parameter values

As you can see, the result is formatted as HTML. We've deliberately returned an HTML tag here so there's a difference between the GET and POST responses. POST is preferable for this type of task because it allows you to play with the returned numbers more easily, and we'll explore it next.

Unlike GET, you can't make POST requests through the browser. Postman (or any other tool) is mandatory. Once the app is open, you'll need to change the request type from GET to POST, enter the URL, `http://localhost:8000/adding`, and under **Body**, select the **Raw** option and **JSON** as the data format. Inside, you'll enter the parameters as specified in the following figure:

Once done, you can click on the **Send** button to submit the request.

Figure 8.5 – The POST endpoint with parameters (Postman)

And that's how you can test your API endpoints through Postman.

In reality, that's not the approach you'll go with. Postman is excellent for testing your APIs, but what if you need to work with the response somehow? Sure, it's a simple number now, but we'll be predicting machine learning models in a couple of pages. If that's the case, sending requests through programming languages such as Python is the only viable solution.

You can use the following code snippet to send requests through Python and capture the response:

```
import requests

req = requests.post(
```

```
    url='http://localhost:8000/adding',
    json={'num1': 3, 'num2': 5}
)
res = req.content

print(res)
```

If you were to run this code now, here's what you would see as the output:

```
b'{"result":8}\n'
```

Figure 8.6 – The POST endpoint with parameters (Python)

This is essentially a string, so some conversion to JSON will be mandatory before you can work with the returned value. More on that later, in *Chapter 9, Using the Deployed TPOT Model in Production.*

So far, you've seen how the `Flask` library can be used to develop both web applications and web services (APIs). It's a good first option, but there's a better approach if you're only interested in building APIs – `Flask-RESTful`. Let's explore it next.

3. You already have `Flask-RESTful` installed. The syntax when using it is a bit different. It uses the **Object-Oriented Programming (OOP)** paradigm for declaring endpoints. In a nutshell, you'll need as many classes as there are distinct endpoints. Each endpoint class can contain one or multiple methods, such as `get()`, `post()`, and `put()`, which represent what happens when a request of a particular type is made.

    All API classes inherit from the `Flask-RESTful.Resource` class and each endpoint must be manually bound to a specific URL string via the `add_resource()` method.

    To summarize, we'll have the `Add` class, which has two methods: `get()` and `post()`. All of the logic inside these methods is identical to what we had earlier, with a single exception – we won't return HTML anywhere.

    Here's the entire code snippet:

```
from flask import Flask, request, jsonify
from flask_restful import Resource, Api

app = Flask(__name__)
api = Api(app)
```

```
class Adding(Resource):
    @staticmethod
    def get():
        num1 = int(request.args.get('num1'))
        num2 = int(request.args.get('num2'))
        return num1 + num2

    @staticmethod
    def post():
        data = request.get_json()
        num1 = data['num1']
        num2 = data['num2']
        return jsonify({'result': num1 + num2})

api.add_resource(Adding, '/adding')


if __name__ == '__main__':
    app.run(host='0.0.0.0', port=8000)
```

Everything inside the Adding class is available on the /adding endpoint.

As you can see, the API will run on a different port, just to easily distinguish between this API and the previous one.

If you were to open http://localhost:8000/adding now, you'd see the following message:

```
{"message": "Internal Server Error"}
```

Figure 8.7 – Flask-RESTful GET without parameters

We now have the same error as with the default `Flask` API, and the reason is that no parameter values were given in the URL. If you were to change it and call `http://localhost:8000/adding?num1=5&num2=10`, you'd see the following in your browser window:

15

Figure 8.8 – Flask-RESTful GET with parameters

As mentioned earlier, communicating with an API straight from the browser is not considered to be a good practice, but you can still do it with GET request types. You're better off using a tool such as Postman, and you already know how to do so.

As for the POST method, you can call the same URL as previously, `http://localhost:8000/adding`, and pass the parameters as JSON in the request body. Here's how you can do so with Postman:

http://localhost:8500/add

Figure 8.9 – Flask-RESTful POST with Postman

You can do the same through Python, but you should already know how to do so by now.

You now know the basics of REST API development with Python, `Flask`, and `Flask-RESTful`. This was a relatively quick, hands-on section that served as a primer for what's about to follow. In the next section, we'll go over a couple of best practices for deploying machine learning models, and in the last two sections, we'll explore how you can train and deploy models to localhost and the cloud, respectively.

Best practices for deploying automated models

The deployment of automated models is more or less identical to the deployment of your normal machine learning models. It boils down to training the model first and then saving the model in some format. In the case of normal machine learning models, you could easily save the model to a `.model` or `.h5` file. There's no reason not to do the same with TPOT models.

If you remember from previous chapters, TPOT can export the best pipeline to a Python file so this pipeline can be used to train the model if it isn't trained already, and the model can be saved afterward. If the model is already trained, only the prediction is obtained.

The check for whether a model has been trained or not can be made by checking whether a file exists or not. If a model file exists, we can assume the model was trained, so we can load it and make a prediction. Otherwise, the model should be trained and saved first, and only then can the prediction be made.

It's also a good idea to use a POST request type when connecting to a machine learning API. It's a better option than GET because parameter values aren't passed directly in the URL. As you might know, parameter values can be sensitive, so it's a good idea to hide them whenever possible.

For example, maybe you need to authenticate with the API before making predictions. It's easy to understand why sending your username and password credentials directly in a URL isn't a good idea. POST has you covered, and the rest of the chapter will make good use of it.

In a nutshell, you should always check whether the model is trained before making predictions and train it if required. The other take-home point is that POST is better than GET in our case. You now know a couple of basic best practices for deploying machine learning models. In the next section, we will train and deploy the model to localhost.

Deploying machine learning models to localhost

We'll have to train a model before we can deploy it. You already know everything about training with TPOT, so we won't spend too much time here. The goal is to train a simple Iris classifier and export the predictive functionality somehow. Let's go through the process step by step:

1. As always, the first step is to load in the libraries and the dataset. You can use the following piece of code to do so:

```
import pandas as pd

df = pd.read_csv('data/iris.csv')
df.head()
```

This is what the first few rows look like:

	sepal_length	sepal_width	petal_length	petal_width	species
0	5.1	3.5	1.4	0.2	setosa
1	4.9	3.0	1.4	0.2	setosa
2	4.7	3.2	1.3	0.2	setosa
3	4.6	3.1	1.5	0.2	setosa
4	5.0	3.6	1.4	0.2	setosa

Figure 8.10 – The first few rows of the Iris dataset

2. The next step is to separate the features from the target variable. This time, we won't split the dataset into training and testing subsets, as we don't intend to evaluate the model's performance. In other words, we know the model performs well, and now we want to retrain it on the entire dataset. Also, string values in the target variables will be remapped to their integer representation as follows:

a) Setosa – 0

b) Virginica – 1

c) Versicolor – 2

The following lines of code do everything that was described:

```
X = df.drop('species', axis=1)
y = df['species']
y = y.replace({'setosa': 0, 'virginica': 1, 'versicolor': 2})
y
```

Here's what the target variable looks like now:

```
0      0
1      0
2      0
3      0
4      0
      ..
145    1
146    1
147    1
148    1
149    1
Name: species, Length: 150, dtype: int64
```

Figure 8.11 – The target variable after value remapping

3. Next stop – model training. We'll train the model with TPOT for 15 minutes. This part should be familiar to you, as we're not using any parameters that weren't used or described in previous chapters.

The following piece of code will train the model on the entire dataset:

```
from tpot import TPOTClassifier


clf = TPOTClassifier(
    scoring='accuracy',
    max_time_mins=15,
    random_state=42,
    verbosity=2
)


clf.fit(X, y)
```

You'll see many outputs during the training process, but it shouldn't take too long to achieve 100% accuracy, as shown in the following figure:

```
Generation 1 - Current best internal CV score: 0.9866666666666667

Generation 2 - Current best internal CV score: 0.9866666666666667

Generation 3 - Current best internal CV score: 0.9866666666666667

Generation 4 - Current best internal CV score: 0.9866666666666667

Generation 5 - Current best internal CV score: 0.9866666666666667

Generation 6 - Current best internal CV score: 0.9866666666666667

Generation 7 - Current best internal CV score: 0.9866666666666667

Generation 8 - Current best internal CV score: 1.0

Generation 9 - Current best internal CV score: 1.0

Generation 10 - Current best internal CV score: 1.0
```

Figure 8.12 – TPOT training process

How many generations will pass in the 15-minute time frame depends on your hardware, but once finished, you should see something similar to this:

```
15.02 minutes have elapsed. TPOT will close down.
TPOT closed during evaluation in one generation.
WARNING: TPOT may not provide a good pipeline if TPOT is stopped/interrupted in a early generation.

TPOT closed prematurely. Will use the current best pipeline.

Best pipeline: KNeighborsClassifier(MultinomialNB(input_matrix, alpha=100.0, fit_prior=False), n_neighbors=10, p=2, weights=distance)
TPOTClassifier(config_dict=None, crossover_rate=0.1, cv=5,
               disable_update_check=False, early_stop=None, generations=100,
               log_file=None, max_eval_time_mins=5, max_time_mins=15,
               memory=None, mutation_rate=0.9, n_jobs=1, offspring_size=None,
               periodic_checkpoint_folder=None, population_size=100,
               random_state=42, scoring='accuracy', subsample=1.0,
               template=None, use_dask=False, verbosity=2, warm_start=False)
```

Figure 8.13 – TPOT output after training

4. Once the training process is complete, you'll have access to the `fitted_pipeline_` property:

```
clf.fitted_pipeline_
```

It's a pipeline object that can be exported for later use. Here's what it should look like (keep in mind that you could see something different on your machine):

```
Pipeline(memory=None,
        steps=[('stackingestimator',
            StackingEstimator(estimator=MultinomialNB(alpha=100.0,
                                                class_prior=None,
                                                fit_prior=False))),
            ('kneighborsclassifier',
            KNeighborsClassifier(algorithm='auto', leaf_size=30,
                                metric='minkowski', metric_params=None,
                                n_jobs=None, n_neighbors=10, p=2,
                                weights='distance')))],
        verbose=False)
```

Figure 8.14 – TPOT fitted pipeline

5. To demonstrate how this pipeline works, please take a look at the following code snippet. It calls the `predict()` function of the `fitted_pipeline_` property with a 2D array of input data, representing a single flower species:

```
clf.fitted_pipeline_.predict([[5.1, 3.5, 0.2, 3.4]])
```

The results are displayed in the following figure:

```
array([0])
```

Figure 8.15 – TPOT prediction

Remember our remapping strategy from a couple of pages ago? 0 indicates that this species was classified as `setosa`.

6. The final step we have to do is save the predictive capabilities of this model to a file. The `joblib` library makes this step easy to do, as you just have to call the `dump()` function to save the model and the `load()` function to load the model.

    Here's a quick demonstration. The goal is to save the `fitted_pipeline_` property to a file called `iris.model`. You can use the following code to do so:

```
import joblib
```

```
joblib.dump(clf.fitted_pipeline_, 'iris.model')
```

And that's all there is to it! You'll see the following output once the model is saved to a file:

```
['iris.model']
```

Figure 8.16 – Saving TPOT models

Just to verify that the model will still work, you can use the `load()` function to load the model in a new variable:

```
loaded_model = joblib.load('iris.model')
loaded_model.predict([[5.1, 3.5, 0.2, 3.4]])
```

The output of the preceding code is shown in the following figure:

```
array([0])
```

Figure 8.17 – Prediction of a saved model

And that's how easy it is to save machine learning models for later use. We now have everything needed to deploy this model, so let's do that next.

The model deployment process will be quite similar to what we've done previously with `Flask` and `Flask-RESTful`. Before proceeding to the step-by-step guide, you should create a directory for your API with the following directory/file structure:

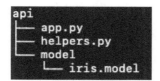

Figure 8.18 – API directory structure

As you can see, the root folder is called `api` and it has two Python files inside – `app.py` and `helpers.py`. The folder also has another folder for storing the model trained previously.

Let's build the API one step at a time next:

1. Let's start with the `helpers.py` file. The goal of this Python file is to remove all calculations and data operations from `app.py`. The ladder is used only to declare and manage the API itself, and everything else is performed elsewhere.

    The `helpers.py` file will have two functions – `int_to_species(in_species)` and `predict_single(model, X)`.

    The goal of the first function is to reverse our previously declared mappings and to return the actual flower species name given the integer representation. Here's a concrete list of strings returned when given the integer input:

    a) 0 – `setosa`

    b) 1 – `virginica`

    c) 2 – `versicolor`

If some other number is passed, an empty string is returned. You can find the code for this function as follows:

```
def int_to_species(in_species):
    if in_species == 0:
        return 'setosa'
    if in_species == 1:
        return 'virginica'
    if in_species == 2:
        return 'versicolor'
```

On to the `predict_single(model, X)` function now. It aims to return a prediction and its probability given the model and a list of input values. The function also makes the following checks:

a) Is X a list? If not, raise an exception.

b) Does X have four items (sepal length, sepal width, petal length, and petal width)? If not, raise an exception.

These checks are required because we don't want bad or misformatted data going in our model and crashing the API.

If all of the checks pass, the prediction and probability are returned to the user as a dictionary, alongside the entered data for each parameter. Here's how to implement this function:

```
def predict_single(model, X):
    if type(X) is not list:
        raise Exception('X must be of list data type!')
    if len(X) != 4:
        raise Exception('X must contain 4 values - \
sepal_length, sepal_width, petal_length, petal_width')
    prediction = model.predict([X])[0]
    prediction_probability =\
model.predict_proba([X])[0][prediction]
    return {
        'In_SepalLength': X[0],
        'In_SepalWidth': X[1],
        'In_PetalLength': X[2],
        'In_PetalWidth': X[3],
```

```
        'Prediction': int_to_species(prediction),
        'Probability': prediction_probability
    }
```

Here's one example of calling the `predict_single()` function:

```
predict_single(
    model=joblib.load('api/model/iris.model'),
    X=[5.1, 3.5, 0.2, 3.4]
)
```

The results are shown in the following figure:

```
{'In_SepalLength': 5.1,
 'In_SepalWidth': 3.5,
 'In_PetalLength': 0.2,
 'In_PetalWidth': 3.4,
 'Prediction': 'setosa',
 'Probability': 1.0}
```

Figure 8.19 – Results of calling the predict_single() function

2. On to `app.py` now. If you have been following along from the beginning of this chapter, coding out this file will be a piece of cake. The goal is to have the model loaded at all times and to trigger the `post()` method of the `PredictSpecies` class when a `/predict` endpoint is called. You'll have to implement both the class and the method yourself.

    The user has to pass input data as JSON. To be more precise, every flower measurement value is passed separately, so the user will have to specify values for four parameters in total.

    If everything goes well, the `predict_single()` function from `helpers.py` is called, and the results are returned to the user.

    Let's take a look at the implementation of `app.py`:

```
import joblib
import warnings
from flask import Flask, request, jsonify
from flask_restful import Resource, Api
from helpers import predict_single
warnings.filterwarnings('ignore')
```

```
app = Flask(__name__)
api = Api(app)
model = joblib.load('model/iris.model')


class PredictSpecies(Resource):
    @staticmethod
    def post():
        user_input = request.get_json()
        sl = user_input['SepalLength']
        sw = user_input['SepalWidth']
        pl = user_input['PetalLength']
        pw = user_input['PetalWidth']

        prediction =\
predict_single(model=model, X=[sl, sw, pl, pw])
        return jsonify(prediction)


api.add_resource(PredictSpecies, '/predict')


if __name__ == '__main__':
    app.run(host='0.0.0.0', port=8000)
```

3.   You now have everything needed to run the API. You can do so the same way
     that you did with the previous APIs, and that is by executing the following line
     in the terminal:

```
> python app.py
```

If everything went well, you'll get the following message:

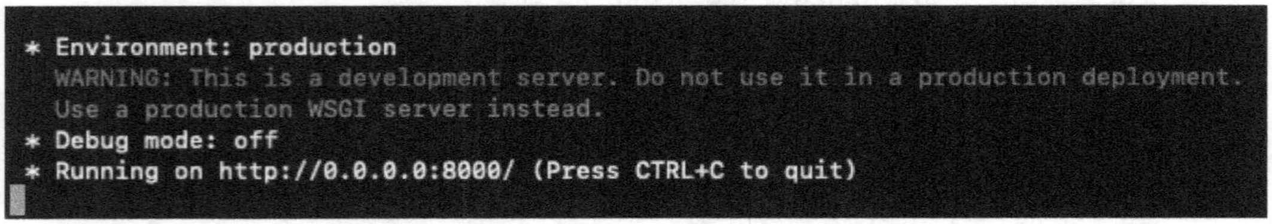

```
* Environment: production
  WARNING: This is a development server. Do not use it in a production deployment.
  Use a production WSGI server instead.
* Debug mode: off
* Running on http://0.0.0.0:8000/ (Press CTRL+C to quit)
```

Figure 8.20 – Running the API

4. The API is now running on `http://localhost:8000`. We'll use the Postman application to test the API.

    Here's the first example:

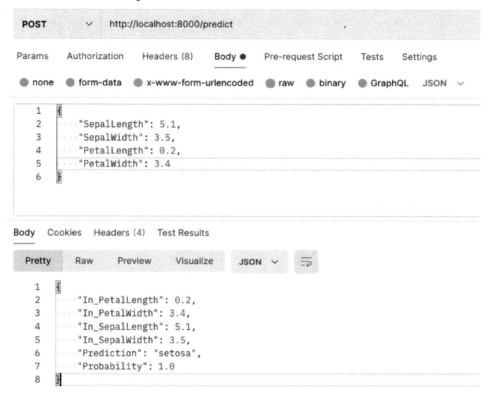

Figure 8.21 – API testing example 1

As you can see, the model is 100% confident that this species belongs to the `setosa` class. Let's try another one:

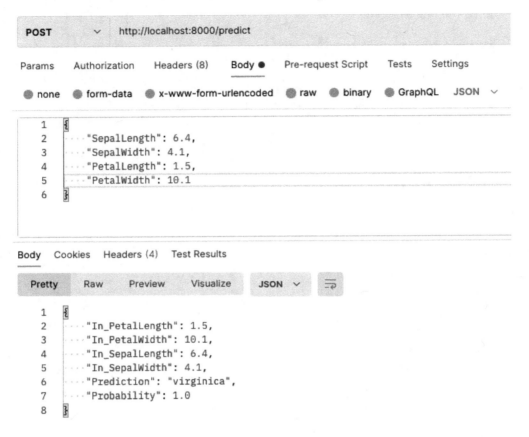

Figure 8.22 – API testing example 2

This has an identical confidence level but a different prediction class. Let's mix things up a beat and pass values a bit different than anything in the training set:

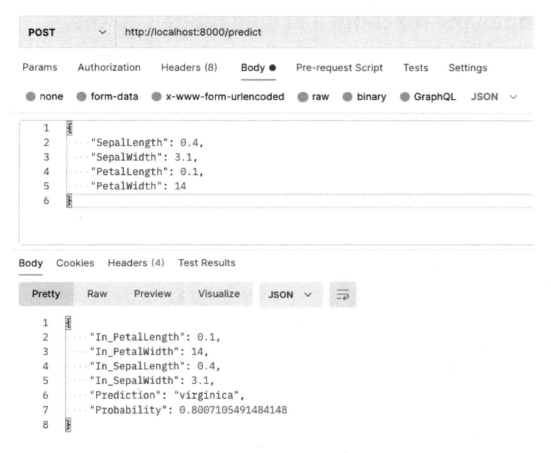

Figure 8.23 – API testing example 3

As you can see, the model isn't 100% confident this time, as the input data is a lot different from the data seen at training.

And there you have it – TPOT model deployment to localhost! The only thing left to do in this chapter is to bring the model to the cloud and make it accessible from anywhere. Let's do that next.

Deploying machine learning models to the cloud

Cloud deployment of machine learning models means creating a cloud virtual machine, transferring our API to it, and running it. It's a tedious process that gets easier with repetition since a lot of steps are involved. If you follow every step precisely from this section, everything will work fine. Just make sure not to miss any minor details:

1. To start, head over to `https://portal.aws.amazon.com/billing/signup#/start` and create an account (assuming you don't already have one). Here's what the website currently looks like (as of February 2021):

Sign up for AWS

Explore Free Tier products with a new AWS account.

To learn more, visit aws.amazon.com/free.

Email address
You will use this email address to sign in to your new AWS account.

⚠ An email address is required.

Password

⚠ A password is required.

Confirm password

AWS account name
Choose a name for your account. You can change this name in your account settings after you sign up.

Continue (step 1 of 5)

Sign in to an existing AWS account

Figure 8.24 – AWS registration website

The registration process will take some time, and you will have to enter your credit card information. Don't worry; we'll create entirely free virtual machine instances so you won't be charged a dime.

2. Once the registration process is complete, click on the **Launch a virtual machine With EC2** option:

Build a solution
Get started with simple wizards and automated

Launch a virtual machine
With EC2

2-3 minutes

Figure 8.25 – EC2 virtual machine creation

We'll create an Ubuntu 20.04 instance. If it's not immediately visible, type `ubuntu` in the search bar:

Figure 8.26 – Ubuntu Server 20.04

Once you click on **Select**, you'll have to specify the type. Make sure to select the free version if you don't want to get charged:

Step 2: Choose an Instance Type

Amazon EC2 provides a wide selection of instance types optimized to fit different use cases. Instances are virtual servers that can run applications. They have varying combinations of CPU, memory, storage, and networking capacity, and give you the flexibility to choose the appropriate mix of resources for your applications. Learn more about instance types and how they can meet your computing needs.

Filter by: All instance families ⌄   Current generation ⌄   Show/Hide Columns

Currently selected: t2.micro (- ECUs, 1 vCPUs, 2.5 GHz, -, 1 GiB memory, EBS only)

	Family	Type	vCPUs ⓘ	Memory (GiB)	Instance Storage (GB) ⓘ	EBS-Optimized Available ⓘ	Network Performance ⓘ	IPv6 Support ⓘ
◯	t2	t2.nano	1	0.5	EBS only	-	Low to Moderate	Yes
■	t2	t2.micro Free tier eligible	1	1	EBS only	-	Low to Moderate	Yes

Figure 2.27 – Ubuntu instance type

Next, click on the **Review and Launch** button. You'll be taken to the following screen:

▼ AMI Details   Edit AMI

🔘 **Ubuntu Server 20.04 LTS (HVM), SSD Volume Type - ami-0996d3051b72b5b2c**

Free tier eligible   Ubuntu Server 20.04 LTS (HVM),EBS General Purpose (SSD) Volume Type. Support available from Canonical (http://www.ubuntu.com/cloud/services).

Root Device Type: ebs   Virtualization type: hvm

▼ Instance Type   Edit instance type

Instance Type	ECUs	vCPUs	Memory (GiB)	Instance Storage (GB)	EBS-Optimized Available	Network Performance
t2.micro	-	1	1	EBS only	-	Low to Moderate

▼ Security Groups   Edit security groups

Security group name   launch-wizard-1
Description   launch-wizard-1 created 2021-02-23T11:19:35.862+01:00

Cancel   Previous   Launch

Figure 2.28 – Ubuntu instance confirmation

Once you click on **Launch**, the following window will appear. Make sure to select the identical options, but the key pair name is up to you:

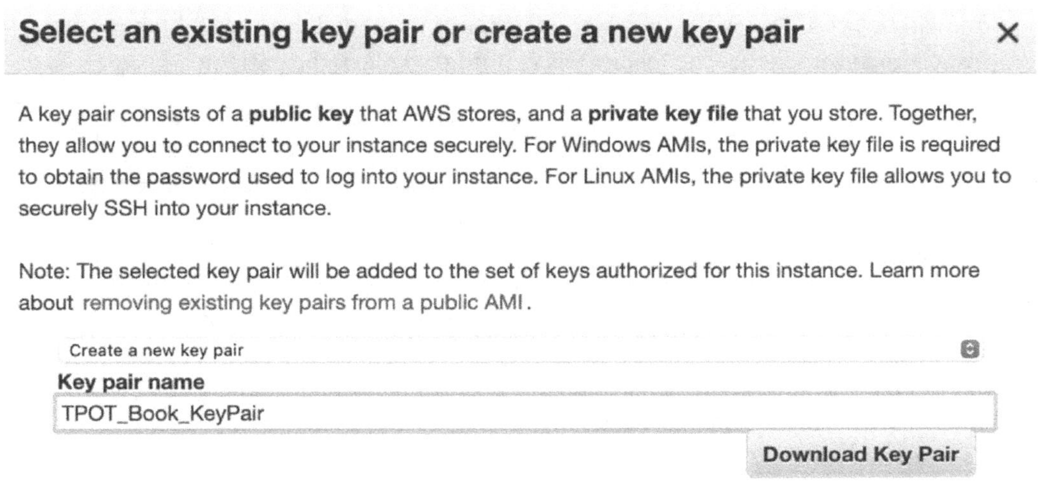

Select an existing key pair or create a new key pair   ✕

A key pair consists of a **public key** that AWS stores, and a **private key file** that you store. Together, they allow you to connect to your instance securely. For Windows AMIs, the private key file is required to obtain the password used to log into your instance. For Linux AMIs, the private key file allows you to securely SSH into your instance.

Note: The selected key pair will be added to the set of keys authorized for this instance. Learn more about removing existing key pairs from a public AMI.

Create a new key pair   🔁

Key pair name

TPOT_Book_KeyPair

Download Key Pair

Figure 8.29 – Ubuntu key pair

Click on the **Download Key Pair** button after you enter the details. After the download is complete, you'll be able to click on the **Launch Instances** button:

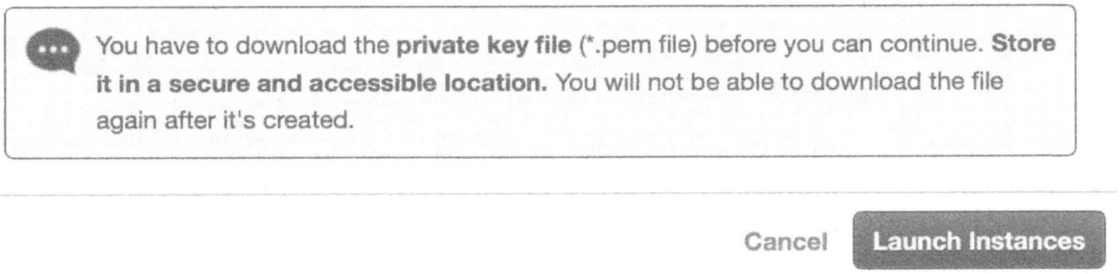

💬 You have to download the **private key file** (*.pem file) before you can continue. **Store it in a secure and accessible location.** You will not be able to download the file again after it's created.

Cancel   **Launch Instances**

Figure 8.30 – Launching the Ubuntu instance

Finally, after everything is done, you can click on the **View Instances** button:

View Instances

Figure 8.31 – View Instances

You will be presented with your created instance immediately. It might take some time before you see that the instance is running, so please be patient:

Figure 8.32 – Running instance

3. To obtain the connection parameter, click on the instance row and select the **Actions | Connect** option, as shown:

Figure 8.33 – Obtaining instance connection parameters

On the screen that follows, you'll see the instance name and dedicated IP address. Here's what it should look like:

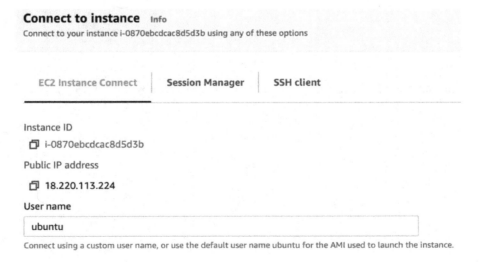

Figure 8.34 – Instance connection parameters

You now have almost everything needed to connect to the instance and transfer the API code. The last piece of the puzzle is the conversion of the key pair `.pem` file to a `.ppk` file. If you're on Windows, this conversion can be made by downloading the *PuTTYgen* app: `https://www.puttygen.com`. If you're on Mac, you can install it directly through Homebrew:

```
> brew install putty
```

Once installed, execute the following terminal command to make the conversion:

```
> puttygen TPOT_Book_KeyPair.pem -o TPOT_Book_PrivateKey.
ppk
```

Once the conversion is made, you can open a file transfer tool such as *FileZilla* (`https://filezilla-project.org`) and establish a connection as shown:

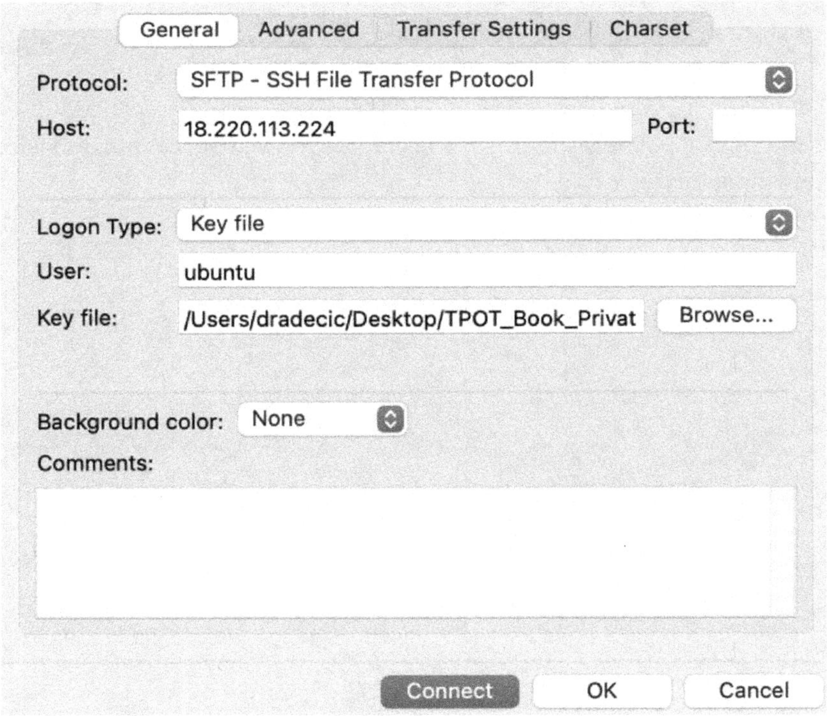

Figure 8.35 – FileZilla connection

Just make sure to point to the `.ppk` file in the **Key file** option. After the **Connect** button is pressed, you'll see the following:

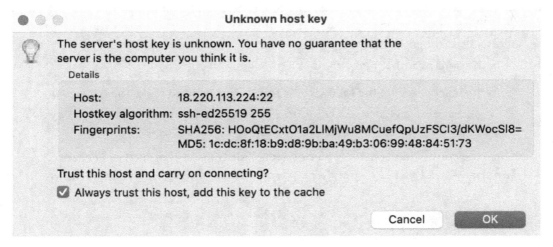

Figure 8.36 – FileZilla host key

Just press **OK** and you're good to go. The connection was successful, as visible from the following figure:

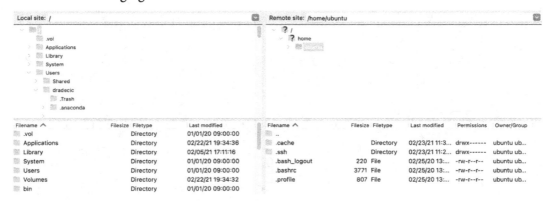

Figure 8.37 – FileZilla successful connection

You can now drag the **api** folder to the **ubuntu** folder on the remote virtual machine, as shown here:

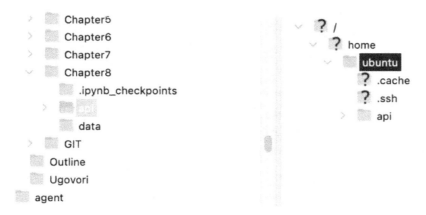

Figure 8.38 – Transferring API data to a remote virtual machine

Before further configuring and starting the API, let's explore how you can obtain the connection through the terminal.

4. Open up a new terminal window right where your `.pem` file is stored. Once there, execute the following command to change the permission:

```
> chmod 400 TPOT_Book_KeyPair.pem
```

Now you have everything needed to establish a secure connection. The following command establishes a connection on my machine:

```
> ssh -i "TPOT_Book_KeyPair.pem" ubuntu@ec2-18-220-113-224.us-east-2.compute.amazonaws.com
```

Please make sure to replace TPOT_Book_KeyPair.pem with your filename and also make sure to write your instance name after ubuntu@. If you did everything correctly, you should see the following in your terminal:

```
[ubuntu@ip-172-31-33-179:~$ ls
api
```

Figure 8.39 – Accessing the Ubuntu virtual machine through the terminal

As you can see, the /api folder is visible to use from the virtual machine.

While we're here, let's configure our Python environment. We need to install `pip`, but to do so, we first have to update the Linux repositories. The following command does both:

```
> sudo apt-get update && sudo apt-get install python3-pip
```

Finally, let's install every required library. Here's the command for doing so within a virtual environment:

```
> pip3 install virtualenv
> virtualenv tpotapi_env
> source tpotapi_env/bin/activate
> pip3 install joblib flask flask-restful sklearn tpot
```

There are a few steps you still need to complete before starting the API, such as managing security.

5. If you were to run the API now, no errors would be raised, but you wouldn't be able to access the API in any way. That's because we need to "fix" a couple of permissions first. Put simply, our API needs to be accessible from anywhere, and it isn't by default.

To start, navigate to **Network & Security | Security Groups** on the sidebar:

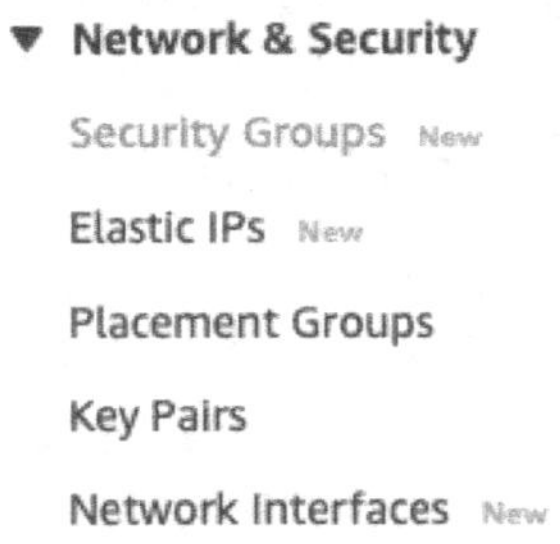

Figure 8.40 – Security Groups

You should see the **Create security group** button in the top-right corner of the browser window:

Figure 8.41 – The Create security group button

Once the new window pops up, you'll have to specify a couple of things. The **Security group name** and **Description** fields are entirely arbitrary. On the other hand, the **Inbound rules** group isn't arbitrary. You'll have to add a new rule with the following options:

a) **Type: All traffic**

b) **Source: Anywhere**

Refer to the following figure for more information:

Create security group Info

A security group acts as a virtual firewall for your instance to control inbound and outbound traffic. To create a new security group, complete the fields below.

Basic details

Security group name Info

| FullAccess |

Name cannot be edited after creation.

Description Info

| FullAccess |

VPC Info

| vpc-1417977f   ▼ |

Inbound rules Info

Type Info	Protocol Info	Port range Info	Source Info		Description - optional Info	
All traffic ▼	All	All	Anywhere ▼	🔍		Delete
				0.0.0.0/0 ✕ ::/0 ✕		

| Add rule |

Figure 8.42 – Creating a security group

After specifying the correct values, you have to scroll down to the end of the screen and click on the **Create security group** option:

Figure 8.43 – Verifying a security group

We're not done yet. The next step is to go to the **Network & Security | Network Interfaces** option on the sidebar:

Figure 8.44 – The Network Interfaces option

Once there, right-click the only available network interface (assuming this is your first time in the AWS console) and select the **Change security groups** option:

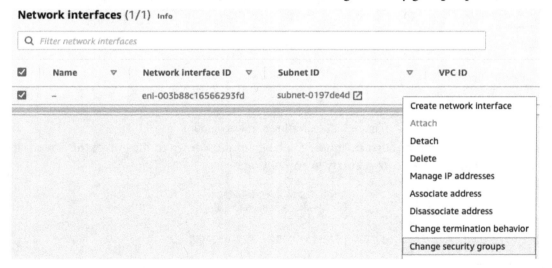

Figure 8.45 – Changing security groups

The whole point of this is to assign the "access from anywhere" rule to our virtual machine. Once a window pops up, select the previously declared security group from the dropdown of options:

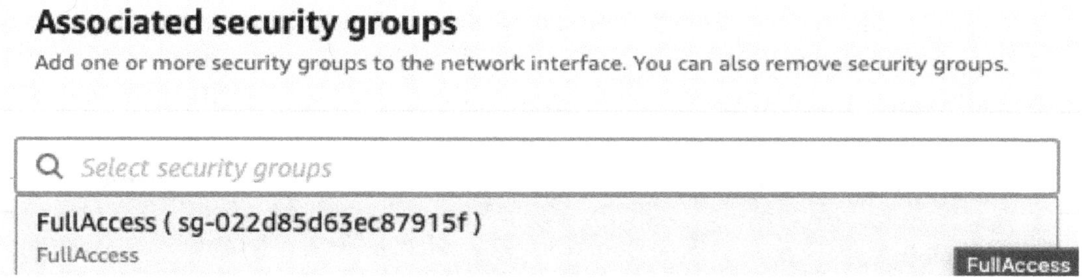

Associated security groups

Add one or more security groups to the network interface. You can also remove security groups.

🔍 *Select security groups*

FullAccess (sg-022d85d63ec87915f)
FullAccess   `FullAccess`

Figure 8.46 – Selecting the security group

Once added, click on the **Save** button to save this association:

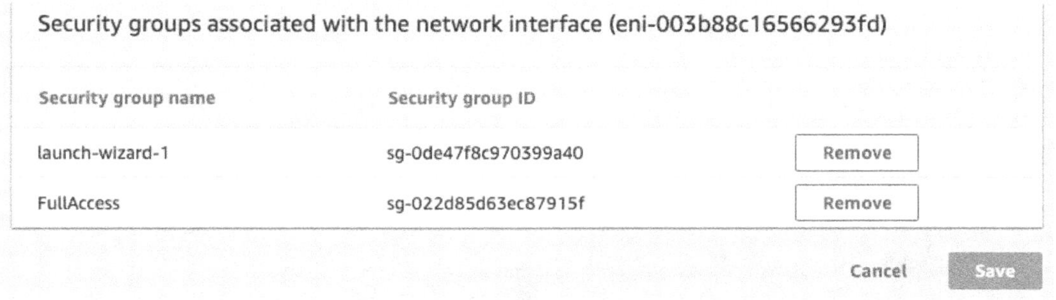

Security groups associated with the network interface (eni-003b88c16566293fd)

Security group name	Security group ID	
launch-wizard-1	sg-0de47f8c970399a40	Remove
FullAccess	sg-022d85d63ec87915f	Remove

Cancel   Save

Figure 8.47 – Saving security associations

6. Configuring our virtual machine was quite a lengthy process, but you can now finally start the `Flask` application (REST API). To do so, navigate to the `/api` folder and execute the following:

```
> python3 app.py
```

You should see the following, by now a familiar message:

```
ubuntu@ip-172-31-33-179:~/api$ python3 app.py
 * Serving Flask app "app" (lazy loading)
 * Environment: production
   WARNING: This is a development server. Do not use it in a production deployment.
   Use a production WSGI server instead.
 * Debug mode: off
 * Running on http://0.0.0.0:8000/ (Press CTRL+C to quit)
```

Figure 8.48 – Starting the REST API through the terminal

And that's it! The API is now running, and we can test whether it works properly.

7. Before making a request through Postman, we first need to find the complete URL for our remote virtual machine. You can find yours by right-clicking the instance under **Instances** and clicking on the **Connect** option. There, you'll see the **SSH client** tab:

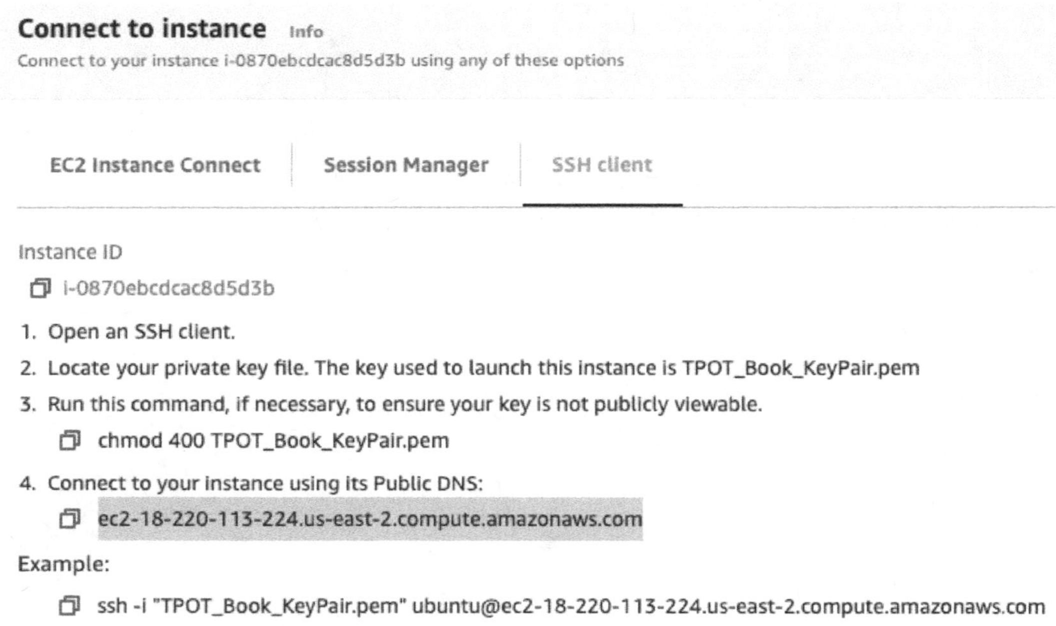

Connect to instance Info

Connect to your instance i-0870ebcdcac8d5d3b using any of these options

| EC2 Instance Connect | Session Manager | SSH client |

Instance ID

⧉ i-0870ebcdcac8d5d3b

1. Open an SSH client.

2. Locate your private key file. The key used to launch this instance is TPOT_Book_KeyPair.pem

3. Run this command, if necessary, to ensure your key is not publicly viewable.

⧉ chmod 400 TPOT_Book_KeyPair.pem

4. Connect to your instance using its Public DNS:

⧉ ec2-18-220-113-224.us-east-2.compute.amazonaws.com

Example:

⧉ ssh -i "TPOT_Book_KeyPair.pem" ubuntu@ec2-18-220-113-224.us-east-2.compute.amazonaws.com

Figure 8.49 – Virtual machine URL

Now we know the complete URL: `http://ec2-18-220-113-224.us-east-2.compute.amazonaws.com:8000/predict`. The procedure from now is identical to as it was on localhost, as visible in the following figure:

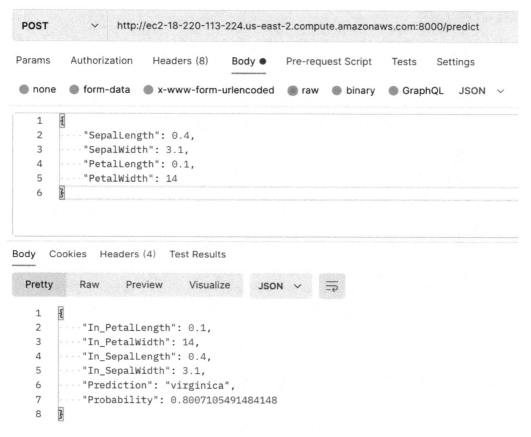

Figure 8.50 – Testing our deployed API

As you can see, the connection went through, and the API returned a response just as it did with the locally deployed version.

And there you have it – the complete procedure for deploying machine learning models to an AWS virtual machine. The process can be quite tedious to do and even tricky if this is your first time. It will get easier as you deploy more and more machine learning models, as the procedure is identical.

You could play around with the permissions if you don't want your API to be accessible from anywhere by anyone, but that's beyond the scope of this book. This chapter is more or less over – great work! What follows is a summary of everything learned and then yet another interesting, hands-on chapter.

Summary

This chapter was the longest one so far and quite intensive with the hands-on tasks. You've hopefully managed to follow along and learned how machine learning models built with TPOT can be deployed – both locally and to the cloud.

You are now capable of deploying any sort of machine learning model built with Python. Besides, you also know how to deploy basic Python web applications, provided that you have the necessary knowledge of frontend technologies, such as HTML, CSS, and JavaScript. We didn't dive into this area, as it's beyond the scope of this book.

In the following chapter, *Chapter 9, Using the Deployed TPOT Model in Production*, you'll learn how to build a basic application around this REST API. To be more precise, you'll learn how to make a simple and decent-looking web interface that predicts flower species based on the input data. But before that, you'll practice making a request to our API with Python.

As always, feel free to study model deployment in more detail, as AWS isn't the only option. There are many cloud providers that offer some sort of a free tier, and AWS is just a single fish in the pond.

Question

1. Why do we need (and want) model deployment?

2. What's the role of REST APIs in model deployment?

3. Name a couple of Python libraries that can be used to deploy machine learning models.

4. What's the difference between the GET and POST request types?

5. What's the general idea behind the `joblib` library in Python?

6. Explain in your own words what a virtual machine is.

7. Which free tool can be used to test REST APIs?

9
Using the Deployed TPOT Model in Production

You've made it to the final chapter – congratulations! So far, you've learned the basics of TPOT by solving classification and regression tasks, how TPOT works with Dask and neural networks, and how to deploy machine learning models both locally and to the cloud.

This chapter will serve as icing on the cake, as you'll learn how to communicate with your deployed models to build something even a 5-year-old could use. To be more precise, you'll learn how to communicate with your API through a notebook environment and a simple GUI web application.

This chapter will cover the following topics:

- Making predictions in a notebook environment
- Developing a simple GUI web application
- Making predictions in a GUI environment

Technical requirements

This is the last chapter in the book, so some prior knowledge is assumed. You need to know how to build basic machine learning models with TPOT to deploy them. It's assumed that your model is deployed to the AWS virtual machine created in *Chapter 8, TPOT Model Deployment*. If that's not the case, please revisit that chapter.

This chapter will be quite code-heavy, so you can refer to the official GitHub repository if you get stuck: `https://github.com/PacktPublishing/Machine-Learning-Automation-with-TPOT/tree/main/Chapter09`.

Making predictions in a notebook environment

If you took a day's (or a few days') break after the previous chapter, it's likely that your connection to the remote virtual machine ended. Because of that, you'll need to reconnect and start the API once again. There are ways to make your API always running, but that's out of the scope of this book. Furthermore, if you've moved the `TPOT_Book_KeyPair.pem` file to some other folder, you'll have to reset the permissions:

1. With that in mind, execute the first command line from the following snippet, only if you have to reset the permissions:

    ```
    > chmod 400 TPOT_Book_KeyPair.pem
    > ssh -i "TPOT_Book_KeyPair.pem" ubuntu@ec2-18-220-113-
    224.us-east-2.compute.amazonaws.com
    > cd api
    > python3 app.py
    ```

2. Your API is running now. The next step is to open a JupyterLab or Jupyter Notebook environment and make a request. You'll need the `requests` library to do so, so here's how to import it:

    ```
    import requests
    ```

    Let's declare a couple of variables next. These will hold the values for the host name, port, and endpoint:

    ```
    HOST ='http://ec2-18-220-113-224.us-east-2.compute.
    amazonaws.com'
    PORT = '8000'
    ENDPOINT = '/predict'
    ```

From there, we can easily concatenate these three variables into a single one to form a URL:

```
URL = f'{HOST}:{PORT}{ENDPOINT}'
URL
```

Here's how it should look:

```
'http://ec2-18-220-113-224.us-east-2.compute.amazonaws.com:8000/predict'
```

Figure 9.1 – URL connection string

Yours will be a bit different due to the difference in the hostname.

3. Next, we'll declare a dictionary that will serve as input data. It will be identical to the data sent in the previous chapter through Postman. Here's the code snippet:

```
in_data = {
    'SepalLength': 0.4,
    'SepalWidth': 3.1,
    'PetalLength': 0.1,
    'PetalWidth': 14
}
```

That's all we need to make a request. Let's do that next.

4. You can use the `post()` function from the `requests` package to make a POST request. Two parameters are required – the URL and the data in JSON format:

```
req = requests.post(url=URL, json=in_data)
req
```

The results are displayed in the following figure:

```
<Response [200]>
```

Figure 9.2 – API response status code

This isn't quite what we were looking for, but a status code of 200 is a good sign that we're on the right track as it indicates a success message. You can access the `content` property of the request to get the actual API response:

```
response = req.content
response
```

Here's how the response looks:

```
b'{"In_PetalLength":0.1,"In_PetalWidth":14,"In_SepalLength":0.4,"In_SepalWidth":3.1,"Prediction":"virginica","Probability":0.8007105491484148}\n'
```

Figure 9.3 – API response as a string

As you can see, the prediction is returned successfully but not in the desired format by default.

5. To change that, you'll need to transform the response string to a JSON object. You can use the `loads()` function from the `json` package to do so:

```
import json

response_json = json.loads(response)
response_json
```

Here are the results:

```
{'In_PetalLength': 0.1,
 'In_PetalWidth': 14,
 'In_SepalLength': 0.4,
 'In_SepalWidth': 3.1,
 'Prediction': 'virginica',
 'Probability': 0.8007105491484148}
```

Figure 9.4 – API response as a JSON object

6. You can access the predicted class (or any other property) just as you would for a normal dictionary object. Here's an example:

```
response_json['Prediction']
```

And here's what's returned:

```
'virginica'
```

Figure 9.5 – API predicted class

And that's essentially how you can obtain predictions from a deployed REST API with Python! In the next section, you'll build a basic interactive web application around this API to make it utterly simple for anyone to use.

Developing a simple GUI web application

This section aims to demonstrate how the `Flask` framework can be used to develop a simple web application. The focus is shifted toward building an application that captures form data, which is then passed to our deployed machine learning API:

1. To start, create the following directory structure:

Figure 9.6 – Web application directory structure

Most of the logic is handled in app.py, and the templates folder is used to store HTML files for the app – more on that in a bit.

2. This time we'll organize the code a bit better, so you'll need to create an additional file for storing the environment variables. Inside the root directory (webapp), create a file called .env – and populate it with the following:

```
SECRET_KEY=SecretKey
HOST=0.0.0.0
PORT=9000
API_ENDPOINT=http://ec2-18-220-113-224.us-east-2.compute.
amazonaws.com:8000/predict
```

Creating a separate file like this is considered to be a best practice when developing any sort of web application.

To use these environment variables, you'll have to install an additional package to your virtual environment:

```
> pip install python-dotenv
```

3. Let's build the basic structure of the application now. Open the app.py file and write the following code:

```
import os
from flask import Flask, render_template
from dotenv import load_dotenv
load_dotenv('.env')

app = Flask(__name__)

@app.route('/')
def index():
    return render_template('index.html')

if __name__ == '__main__':
    app.run(host=os.getenv('HOST'), port=os.
getenv('PORT'))
```

If you were to run the application now, you wouldn't get an error, but nothing would be displayed on the screen. The reason is simple – we haven't handled the `index.html` file yet. Before we do so, let's discuss the only potentially unfamiliar part of code: the `render_template()` function. Put simply, this function will display an HTML file instead of showing merely a string or value returned from the function. There's a way to pass parameters, but more on that later.

4. Onto the `index.html` now – here's the code you can paste inside the file:

```
<!DOCTYPE html>
<html lang="en">
<head>
    <meta charset="UTF-8">
    <meta http-equiv="X-UA-Compatible" content="IE=edge">
    <meta name="viewport" content="width=device-width,
initial-scale=1.0">
    <title>Iris Predictioner</title>
</head>
<body>
    <h1>Welcome to Iris Predictioner</h1>
</body>
</html>
```

Don't worry if you haven't written a line of HTML before – it's a simple markup language. Think of everything you see here as boilerplate. It's what's inside the `<body></body>` tag that we're interested in.

If you were to run your application now, here's how it would look:

Welcome to Iris Predictioner

Figure 9.7 – Iris prediction application (v1)

It's a simple and utterly boring web application, but at least it works.

5. As mentioned before, our web application has to handle form data somehow, so let's start working on that. There's a dedicated package for handling form data with `Flask` called `Flask-WTF`. Here's how you can install it:

```
> pip install flask-wtf
```

Once installed, create a `forms.py` file in the root directory – `/webapp/forms.py`. Let's take a look at the code this file contains and explain it:

```python
from flask_wtf import FlaskForm
from wtforms import FloatField, SubmitField
from wtforms.validators import DataRequired


class IrisForm(FlaskForm):
    sepal_length = FloatField(
        label='Sepal Length', validators=[DataRequired()]
    )
    sepal_width = FloatField(
        label='Sepal Width', validators=[DataRequired()]
    )
    petal_length = FloatField(
        label='Petal Length', validators=[DataRequired()]
    )
    petal_width = FloatField(
        label='Petal Width', validators=[DataRequired()]
    )
    submit = SubmitField(label='Predict')
```

Okay, so what's going on in this file? Put simply, `Flask-WTF` allows us to declare forms for `Flask` applications easily, in a class format. We can use any of the built-in field types and validators. For this simple example, we'll only need float and submit fields (for flower measurements and the submit button). Validation-wise, we only want that no fields are left blank.

That's all you need to do, and `Flask` will take care of the rest.

6. Onto the `app.py` now. A couple of changes are required:

- `Flask-WTF` forms need a secret key configured to work. You can add it by accessing the `.env` file. What you declare as a value is entirely arbitrary.

- Our index route now needs to allow for both POST and GET methods since it will handle forms. Inside the `index()` function, you'll have to instantiate the previously written `IrisForm` class and return relevant results if there are no validation errors once the submit button is clicked.

  You can use the `validate_on_submit()` function to check. If the check is passed, the input data is returned in a heading format (we'll see how to show predictions later). If not, the `index.html` template is returned.

- A call to `render_template()` now passes a parameter to our HTML file – `iris_form`. This gives access to form data to our HTML file. You'll see how to deal with it in a minute.

Here's how your file should look once the changes are made:

```python
import os
from flask import Flask, render_template
from forms import IrisForm
from dotenv import load_dotenv
load_dotenv('.env')

app = Flask(__name__)
app.config['SECRET_KEY'] = os.getenv('SECRET_KEY')

@app.route('/', methods=['GET', 'POST'])
def index():
    iris_form = IrisForm()
    if iris_form.validate_on_submit():
        return f'''
                <h3>
                        Sepal Length: {iris_form.sepal_
length.data}<br>
                        Sepal Width: {iris_form.sepal_width.
data}<br>
                        Petal Length: {iris_form.petal_
length.data}<br>
                        Petal Width: {iris_form.petal_width.
data}
                </h3>
                '''

    return render_template('index.html', iris_form=iris_
form)


if __name__ == '__main__':
```

```
app.run(host=os.getenv('HOST'), port=os.
getenv('PORT'))
```

We're almost there. Let's tweak the index.html file next.

7. index.html is the last file you'll need to tweak to have a working application. The only thing we need inside it is a form that displays the fields declared earlier. It's also mandatory to protect your app from **Cross-Site Request Forgery** (**CSRF**) attacks. To do so, you'll have to place a token before the form fields.

Here's how the HTML file should look:

```html
<!DOCTYPE html>
<html lang="en">
<head>
    <meta charset="UTF-8">
    <meta http-equiv="X-UA-Compatible" content="IE=edge">
    <meta name="viewport" content="width=device-width,
initial-scale=1.0">
    <title>Iris Predictioner</title>
</head>
<body>
    <h1>Welcome to Iris Predictioner</h1>
    <form method="POST" action="{{ url_for('index') }}">
        {{ iris_form.csrf_token }}
        {{ iris_form.sepal_length.label }} {{ iris_form.
sepal_length(size=18) }}
        <br>
        {{ iris_form.sepal_width.label }} {{ iris_form.
sepal_width(size=18) }}
        <br>
        {{ iris_form.petal_length.label }} {{ iris_form.
petal_length(size=18) }}
        <br>
        {{ iris_form.petal_width.label }} {{ iris_form.
petal_width(size=18) }}
        <br>
        <input type="submit" value="Predict">
    </form>
</body>
</html>
```

As you can see, to access parameters sent from the Python file, you have to surround the code with double curly brackets.

8. If you were to launch the application now, here's what you'd see on the screen:

Figure 9.8 – Iris prediction application

And that's a frontend for your machine learning application! It's a bit ugly, but we'll style it later. Let's test the functionality first.

We don't want the form submitted if any of the input values are empty. Here's what happens if the button is pressed immediately:

Figure 9.9 – Iris prediction application form validation (1)

Validation test 1 – check. Let's see what happens if only one input field remains empty:

Figure 9.10 – Iris prediction application form validation (2)

The same message occurs, just as you would expect. To conclude, the form can't be submitted if any of the input fields are empty.

To continue, fill out all of the fields, as shown in the following figure:

Figure 9.11 – Iris prediction application form values

If you were to click on the button now, here's the result you'd see:

Sepal Length: 3.3
Sepal Width: 3.5
Petal Length: 3.7
Petal Width: 4.1

Figure 9.12 – Iris prediction application results

So far, everything works, but there's still one step we should do before connecting the application to our Iris prediction API – styling. This step is optional, as the application will still work if you decide to jump to the API connection part immediately.

Setting adequate styles to your Flask application will require a bit of work and refactoring. You'll find the entire list of steps here. Keep in mind that this book assumes no HTML and CSS knowledge. You're free to copy and paste the content of these files but are encouraged to explore further on your own:

1. Let's start with app.py. Instead of returning an H2 tag with input values printed as a single long string, we'll return an HTML template that will show a table. For now, we'll fill out the input data only and set dummy values for prediction and prediction probability.

    Here's how the file should look after the changes:

    ```
    import os
    from flask import Flask, render_template
    from forms import IrisForm
    from dotenv import load_dotenv
    load_dotenv('.env')

    app = Flask(__name__)
    app.config['SECRET_KEY'] = os.getenv('SECRET_KEY')

    @app.route('/', methods=['GET', 'POST'])
    def index():
        iris_form = IrisForm()
        if iris_form.validate_on_submit():
            return render_template(
                'predicted.html',
    ```

```
            sepal_length=iris_form.sepal_length.data,
            sepal_width=iris_form.sepal_width.data,
            petal_length=iris_form.petal_length.data,
            petal_width=iris_form.petal_width.data,
            prediction='Prediction',
            probability=100000
        )
    return render_template('index.html', iris_form=iris_
form)


if __name__ == '__main__':
    app.run(host=os.getenv('HOST'), port=os.
getenv('PORT'))
```

2. Let's create a template file while we're at it. Under /templates, create a predicted.html file. As mentioned earlier, this file will contain a table showing the API response (once we implement it).

Here's how the file should look:

```
<!DOCTYPE html>
<html lang="en">
<head>
    <meta charset="UTF-8">
    <meta http-equiv="X-UA-Compatible" content="IE=edge">
    <meta name="viewport" content="width=device-width,
initial-scale=1.0">
    <link rel="stylesheet" href="{{ url_for('static',
filename='css/main.css') }}">
    <title>Iris Predictioner</title>
</head>
<body>
    <div class="container">
        <h1>Predictions:</h1>
        <table>
            <thead>
              <tr><th>Attribute</th><th>Value</th></tr>
            </thead>
            <tbody>
                <tr><td>Sepal Length</td><td>{{ sepal_
length }}</td></tr>
                <tr><td>Sepal Width</td><td>{{ sepal_width
}}</td></tr>
```

```
                <tr><td>Petal Length</td>td>{{ petal_length
}}</td></tr>
                <tr><td>Petal Width</td><td>{{ petal_width
}}</td></tr>
                <tr><td>Prediction</td><td>{{ prediction
}}</td></tr>
                <tr><td>Probability</td><td>{{ probability
}}</td></tr>
            </tbody>
        </table>
    </div>
</body>
</html>
```

As you can see, we've utilized the power of parameter passing to show the data going in and out of the predictive model. If you're wondering what the deal is with CSS file linking in the document head – don't worry about it for now. We still need to work on one thing before dealing with CSS.

3. Finally, let's reformat index.html. This file will need only minor changes – a couple of CSS classes and a couple of div elements. Here is the entire code snippet for the reformatted version:

```
<!DOCTYPE html>
<html lang="en">
<head>
    <meta charset="UTF-8">
    <meta http-equiv="X-UA-Compatible" content="IE=edge">
    <meta name="viewport" content="width=device-width,
initial-scale=1.0">
    <link rel="stylesheet" href="{{ url_for('static',
filename='css/main.css') }}">
    <title>Iris Predictioner</title>
</head>
<body>
    <div class="container">
        <h1>Welcome to Iris Predictioner</h1>
        <form method="POST" action="{{ url_for('index')
}}">
            {{ iris_form.csrf_token }}
            <div class="single-input">
                {{ iris_form.sepal_length.label }} {{
iris_form.sepal_length(size=18) }}
            </div>
```

```
        <div class="single-input">
                {{ iris_form.sepal_width.label }} {{
iris_form.sepal_width(size=18) }}
        </div>
        <div class="single-input">
                {{ iris_form.petal_length.label }} {{
iris_form.petal_length(size=18) }}
        </div>
        <div class="single-input">
                {{ iris_form.petal_width.label }} {{
iris_form.petal_width(size=18) }}
        </div>
        <input class="btn-submit" type="submit"
value="Predict">
      </form>
    </div>
</body>
</html>
```

4. We're almost there. So far, you've refactored every file that needed refactoring, and now you'll create an additional folder and file. Inside the root directory, create a folder named static. Once created, make an additional folder inside it called css. This folder will contain all stylings for our application. Inside the css folder, create a file called main.css.

To summarize, here's how your directory structure should look once you have created these folders and file:

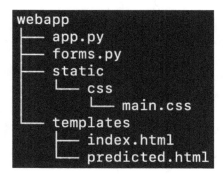

Figure 9.13 – New directory structure

The `main.css` file is already linked to your HTML files, if you remember. If not, please take a look inside the `<head></head>` tag of the HTML files – you'll find the link there.

Here's the code for the entire `main.css` file:

```css
@import url('https://fonts.googleapis.com/
css2?family=Open+Sans:wght@400;600&display=swap');

* { margin: 0; padding: 0; box-sizing: border-box;
    font-family: 'Open Sans', sans-serif; }

body { background-color: #f2f2f2; }

.container { width: 800px; height: 100vh; margin: 0 auto;
    background-color: #ffffff; padding: 0 35px; }

.container > h1 { padding: 35px 0; font-size: 36px;
    font-weight: 600; }

.single-input { display: flex; flex-direction: column;
    margin-bottom: 20px; }

.single-input label { font-weight: 600; }

.single-input label::after { content: ":" }

.single-input input { height: 35px; line-height: 35px;
    padding-left: 10px; }

.btn-submit { width: 100%; height: 35px;
    background-color: #f2f2f2; font-weight: 600;
    cursor: pointer; border: 2px solid #dddddd;
    border-radius: 8px; }

table { font-size: 18px; width: 100%; text-align: left; }
```

And we're done. Let's run the application to see how it looks now.

5. If you were to re-run the application now, you'd see the stylings taking effect. The following figure shows how the input form looks:

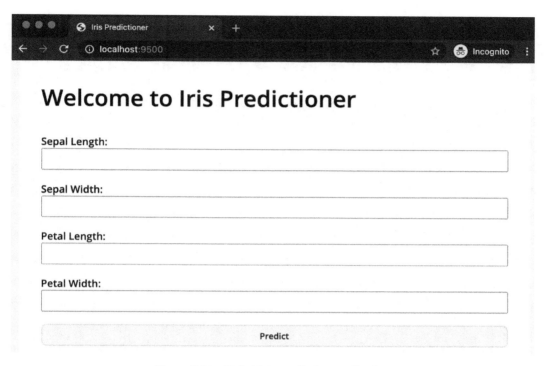

Figure 9.14 – Styled Iris prediction application

The application by no means looks perfect now, but it's at least in a presentable form. Let fill it out as shown in the following figure:

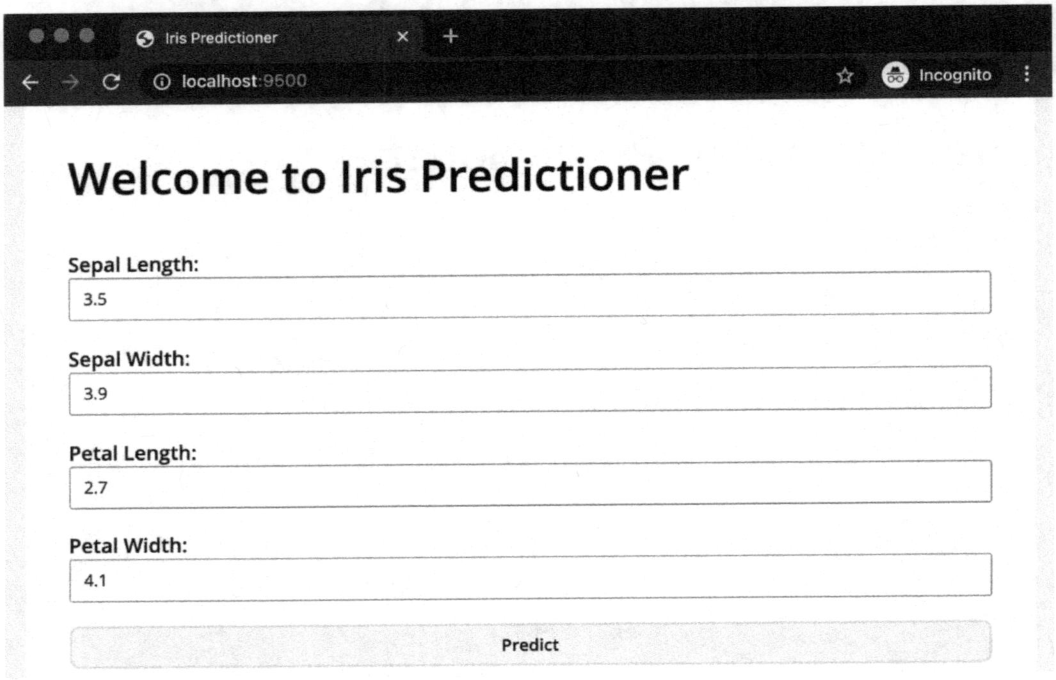

Figure 9.15 – Styled Iris prediction application (2)

And finally, let's click on the **Predict** button to see how the other page looks:

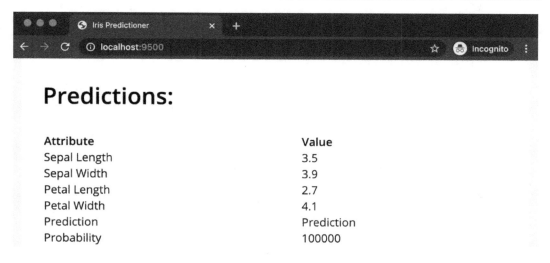

Figure 9.16 – Iris prediction application predictions

Let's call it a day stylings-wise. The application is at a pretty decent point now, but you're free to tweak it further.

And there you have it – how to build and style a `Flask` application built around a machine learning model. The next section will connect the app to our API, making the app fully functional. See you there.

Making predictions in a GUI environment

Welcome to the last section of the book. This section will tie our simple web application to an already-deployed machine learning API. This closely resembles a production environment, where you have one or more machine learning models deployed, and the application development team wants to use them in their application. The only difference is that you're both the data science and application development team.

Once again, we'll have to make a couple of changes to the application structure:

1. Let's start with the simpler part. Inside the root directory, create a Python file called `predictor.py`. This file will hold a single function that implements the logic discussed at the beginning of this chapter when we made predictions in the notebook environment.

Put simply, this function has to make a POST request to the API and return a response in JSON format.

Here's the entire code snippet for the file:

```
import os
import json
import requests
from dotenv import load_dotenv
load_dotenv('.env')


def predict(sepal_length, sepal_width, petal_length,
petal_width):
    URL = os.getenv('API_ENDPOINT')
    req = requests.post(
        url=URL,
        json={
            'SepalLength': sepal_length,
            'SepalWidth': sepal_width,
            'PetalLength': petal_length,
            'PetalWidth': petal_width
        }
    )
    response = json.loads(req.content)
    return response
```

Keep in mind that the URL parameter's value will be different on your machine, so please change it accordingly.

There's no point in further explaining this code snippet, as it is nearly identical to the code you've seen and written before.

2. Let's make a couple of changes in app.py now. We'll import this file and call the predict() function right after the input fields are validated. Once the response is returned, its values are passed as parameters to the corresponding field of the return statement.

Here is the entire code snippet for the app.py file:

```
import os
import numpy as np
from flask import Flask, render_template
from forms import IrisForm
from predictor import predict
from dotenv import load_dotenv
```

```
load_dotenv('.env')

app = Flask(__name__)
app.config['SECRET_KEY'] = os.getenv('SECRET_KEY')

@app.route('/', methods=['GET', 'POST'])
def index():
    iris_form = IrisForm()
    if iris_form.validate_on_submit():
        pred_response = predict(
            sepal_length=iris_form.sepal_length.data,
            sepal_width=iris_form.sepal_width.data,
            petal_length=iris_form.petal_length.data,
            petal_width=iris_form.petal_width.data
        )
        return render_template(
            'predicted.html',
            sepal_length=pred_response['In_PetalLength'],
            sepal_width=pred_response['In_PetalWidth'],
            petal_length=pred_response['In_SepalLength'],
            petal_width=pred_response['In_SepalWidth'],
            prediction=pred_response['Prediction'],
            probability=f"{np.round((pred_
response['Probability'] * 100), 2)}%"
        )
    return render_template('index.html', iris_form=iris_
form)


if __name__ == '__main__':
    app.run(host=os.getenv('HOST'), port=os.
getenv('PORT'))
```

As you can see, the prediction probability is converted to a percentage and rounded to two decimal points. The only reason for doing so is to have nicer-formatted output in the application.

3. And now the fun part – testing. Open the application and enter some data into the form. Here's an example:

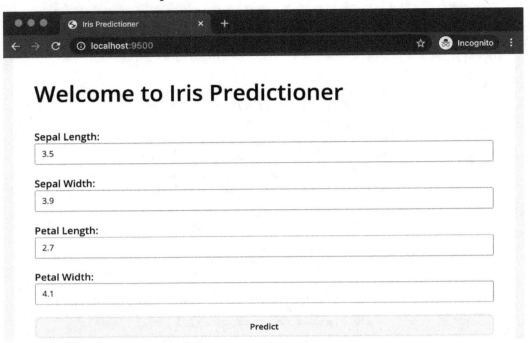

Figure 9.17 – Iris prediction application final test

Once you click on the **Predict** button, you'll see the following results on the screen:

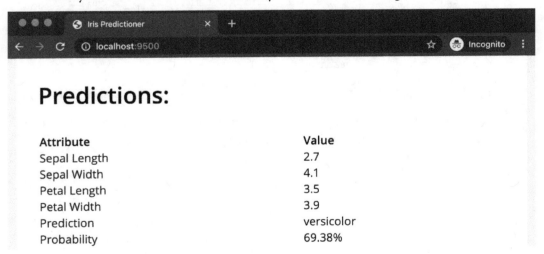

Figure 9.18 – Iris prediction application final results

And there you have it – a complete and fully working GUI web application based on a deployed machine learning model.

Not including the summary that follows, this was the last section of the chapter, but also the entire book. You now know how to deploy machine learning models and build simple web applications around a deployed model. Congratulations!

Summary

This chapter fell into the category of extensive hands-on chapters, but I hope you've managed to follow along. If you have, you've learned a lot – from how to make predictions in a notebook environment to making predictions in a simple and custom-built web application.

Not only that, but you've also completed the entire book. Congratulations! You've learned a lot throughout these nine chapters. We started with the basics of machine learning through basic regression and classification examples, and from there slowly built our knowledge of TPOT. You've also learned how TPOT works with parallel training and with neural networks. But probably the most important new skill you've acquired is model deployment. Without it, your models are useless, as no one can use them to create value.

As always, feel free to explore TPOT and every amazing functionality it has to offer on your own. This book should serve you as a great starting point, as it took you from zero to building web applications around your deployed automated machine learning models in only a couple of hundred pages. Now that's something you can be proud of!

Q&A

1. Which Python library can you use to make requests to deployed REST APIs?

2. In which format is data provided when making a POST request?

3. Name the Flask extension used to build and work with forms.

4. Why is it important to validate web application forms if we're talking about data going into a machine learning model?

5. Can you pass parameters to HTML template files through Flask? If so, how can you display their values in HTML?

6. Explain the process of linking CSS files to Flask applications.

7. Explain why there's no point in leaving machine learning models sitting idle on your PC.

`Packt.com`

Subscribe to our online digital library for full access to over 7,000 books and videos, as well as industry leading tools to help you plan your personal development and advance your career. For more information, please visit our website.

Why subscribe?

- Spend less time learning and more time coding with practical eBooks and Videos from over 4,000 industry professionals

- Improve your learning with Skill Plans built especially for you

- Get a free eBook or video every month

- Fully searchable for easy access to vital information

- Copy and paste, print, and bookmark content

Did you know that Packt offers eBook versions of every book published, with PDF and ePub files available? You can upgrade to the eBook version at `packt.com` and as a print book customer, you are entitled to a discount on the eBook copy. Get in touch with us at `customercare@packtpub.com` for more details.

At `www.packt.com`, you can also read a collection of free technical articles, sign up for a range of free newsletters, and receive exclusive discounts and offers on Packt books and eBooks.

Other Books You May Enjoy

If you enjoyed this book, you may be interested in these other books by Packt:

Automated Machine Learning

Adnan Masood, PhD

ISBN: 978-1-80056-768-9

- Explore AutoML fundamentals, underlying methods, and techniques
- Assess AutoML aspects such as algorithm selection, auto featurization, and hyperparameter tuning in an applied scenario
- Find out the difference between cloud and operations support systems (OSS)
- Implement AutoML in enterprise cloud to deploy ML models and pipelines
- Build explainable AutoML pipelines with transparency

- Understand automated feature engineering and time series forecasting
- Automate data science modeling tasks to implement ML solutions easily and focus on more complex problems

Automated Machine Learning with Microsoft Azure

Dennis Michael Sawyers

ISBN: 978-1-80056-531-9

- Understand how to train classification, regression, and forecasting ML algorithms with Azure AutoML
- Prepare data for Azure AutoML to ensure smooth model training and deployment
- Adjust AutoML configuration settings to make your models as accurate as possible
- Determine when to use a batch-scoring solution versus a real-time scoring solution
- Productionalize your AutoML solution with Azure Machine Learning pipelines
- Create real-time scoring solutions with AutoML and Azure Kubernetes Service
- Discover how to quickly deliver value and earn business trust using AutoML
- Train a large number of AutoML models at once using the AzureML Python SDK

Packt is searching for authors like you

If you're interested in becoming an author for Packt, please visit `authors.packtpub.com` and apply today. We have worked with thousands of developers and tech professionals, just like you, to help them share their insight with the global tech community. You can make a general application, apply for a specific hot topic that we are recruiting an author for, or submit your own idea.

Leave a review - let other readers know what you think

Please share your thoughts on this book with others by leaving a review on the site that you bought it from. If you purchased the book from Amazon, please leave us an honest review on this book's Amazon page. This is vital so that other potential readers can see and use your unbiased opinion to make purchasing decisions, we can understand what our customers think about our products, and our authors can see your feedback on the title that they have worked with Packt to create. It will only take a few minutes of your time, but is valuable to other potential customers, our authors, and Packt. Thank you!

Index

W

X